Early Warning and Multihazard and Disaster Management Systems

Early Warning-Based Multihazard and Disaster Management Systems

Syed Hyder Abbas Musavi

CRC Press
Taylor & Francis Group
Boca Raton London New York

CRC Press is an imprint of the
Taylor & Francis Group, an **informa** business

CRC Press
Taylor & Francis Group
6000 Broken Sound Parkway NW, Suite 300
Boca Raton, FL 33487-2742

First issued in paperback 2023

Library of Congress Cataloging-in-Publication Data

Names: Musavi, Syed Hyder Abbas, author.
Title: Early warning-based multihazard and disaster management systems / Syed Hyder Abbas Musavi.
Description: First edition. | Boca Raton, FL : CRC Press/Taylor & Francis Group, 2020. | Includes bibliographical references.
Identifiers: LCCN 2019014723 | ISBN 9781138391888 (hardback : acid-free paper)
Subjects: LCSH: Emergency management. | Emergency communication systems. | Disaster relief.
Classification: LCC HV551.2 .M875 2020 | DDC 363.34/72--dc23
LC record available at https://lccn.loc.gov/2019014723

Visit the Taylor & Francis Web site at
http://www.taylorandfrancis.com

and the CRC Press Web site at
http://www.crcpress.com

ISBN 13: 978-1-032-65361-7 (pbk)
ISBN 13: 978-1-138-39188-8 (hbk)
ISBN 13: 978-0-429-31990-7 (ebk)

DOI: 10.1201/9780429319907

To my late father and late mother, it's impossible to thank you adequately for everything you've done for me, from loving me unconditionally to raising me as a stable human being, in whom you instilled traditional values and taught your children to celebrate and embrace life. You are my role models.

And to my wife, thanks for all of the wonderful memories and for your continued support and encouragement. A special thank you for showing me that anything is possible with faith, hard work and determination.

For all of my loved ones and family members who have gone on to a better life, you are always close to my heart. For all of my wonderful friends old and new, thanks for always being there for me!

Finally, this dedication would not be complete without a very special thank you to the men and women of the Pakistan Armed Forces, who protect and defend our freedom all over the world. God bless you!

Contents

Preface

A researcher is someone who conducts *research*, that is, an organized and systematic investigation into something. His life is full of craze to serve humanity with new ideas and target solutions that can make lives better. This is especially more evident when a researcher conducts research for human life betterment, and works toward identifying and solving the major issues that create obstacles in a healthy society.

After the events of September 11, 2001, the Kashmir Earthquake of October 2005, the Indian Ocean tsunami of 2004, the Gulf Coast hurricanes of 2005, the terrorist bombings of July 7, 2005 in London, the Karachi Nishtar Park and other suicide blasts in Pakistan and the 2010 floods in all the four provinces of Pakistan, there was a widespread sense in Pakistan and in many other parts of the world that humanity has entered a new and more dangerous era. In this new world of the twenty-first century, it is essential that we anticipate such events (Early Warnings) and their potential impacts. It is impossible to know exactly what form they will take, how severe they will be, or where and when they will occur, but their devastating impacts (pains) can be mitigated through demonstration of extensive strategic planning. The work in this book, therefore, is about the value of a specific area of ICT (Information and Communication Technologies) planning, how Pakistan might make improvements in this specific area and the types and roles these play in disaster and emergency management. Thus, we have proposed a system that demonstrates an interoperable and integrated view of many agencies and organizations in Pakistan that can facilitate and improve the efficacy of routine, day-to-day public safety operations before, during and after a disaster hits a developing country.

The system is based on E-line-cum-state of the art prevailing wireless technology. Specifically, Pakistan's law enforcement agencies (Army, Air Force, Navy and Police), governmental agencies, NADRA, public safety departments, newly proposed emergency response teams and the general public will benefit from this system. The scenarios that are addressed include a "system of systems" (PANs: MANET and VANET; JANs: IEEE 802.16e [also IEEE 802.20], mobile broadband wireless networking and mesh networking technologies; IANs: IEEE 802.11 wireless local area networks; EANs: satellite connectivity and wireless ad hoc networking technologies) that contain operational requirements described using scenarios such as multidisciplined, within a local area (a pre-planned event, a terrorist bomb attack, a local fire incident, etc.) and multidisciplined large-scale regional events (a hurricane, an earthquake, a flood, etc.). Thus, this book discusses novel ideas of applying information and communication technologies such as geospatial maps, which are an essential part of search-and-rescue operations using GIS, about remote sensing using GPS (global positioning system) receivers that allow first responders to analyze zones and locate damaged buildings or injured residents, about images that are captured from aircraft to provide the first comprehensive picture of an event's impact, about road maps that form the basis of evacuation planning and about all other information connected to a location that can be used in emergency management. The analysis was performed using OPNET Modeler to design and develop simulated models corresponding to the multihazard early warning system (with a specific focus on the earthquake early warning system), its various wireless flavors (like WiMAX, Wi-Fi and Zigbee Models also with Satellite Connectivity) and models related to handling issues of communication interoperability and integration between varying public safety organizations, NGOs, law enforcement and investigating agencies, relief organizations using link redundancy principles of wired and wireless telecommunication technologies such as WiMAX, Wi-Fi, Zigbee, Satellite and IP-based broadband connectivity.

This book is divided into eight chapters giving an introduction to the subject design and development of an E-line proactive wireless disaster management and civionics system.

Chapter 1 depicts improvements in the field of disaster engineering that have been elicited following the worldwide web of natural and man-made havocs such as Hurricane Katrina, Pakistan's 2005 earthquake, numerous day-to-day bombing incidents in Pakistan and the 9/11 attacks. The role of an efficient and robust ICT system is essential for reduction of response time and for augmenting meaningful usage of resources during the disaster management phases of relief, response, recovery and rehabilitation. There seems to be an increasing feeling among the general public all over the world for informing them ahead of any disaster to save lives and property. Besides, the shift of the United Nations attention toward disaster mitigation measures also signifies the communications interoperability scheduling. This chapter discusses the role of ICT in disaster management, the worldwide economic burden of disasters and need of a multihazard early warning system.

Chapter 2 discusses disaster management involving four phases. These are preparedness, planning, response and recovery. In the disaster cycle, the steps involved are mitigation (prevention, preparedness and planning), alertness, response and recovery (rehabilitation and reconstruction). Disaster management refers to a collective term encompassing all aspects of planning for and responding to disasters including both the pre- and post-disaster activities. A robust early warning system can reduce the loss of lives and property. Interoperability in a communication network has also remained a key issue during the response phase. This chapter discusses in detail the challenges being faced by an intra-agency communication network and disaster resilience steps.

Chapter 3 briefly describes ICT tools for disaster management, such as open source computer software. Two important tools are usually implemented worldwide in disaster engineering. One is the Hazus Loss Estimation Tool used by the Federal Emergency Management Authority (FEMA) in the United States, and the second is the Sahana Disaster Management Tool. Both these tools are used in the response phase of disaster engineering.

Chapter 4 discusses how to design and develop a generic model of a viable multihazard early warning system for developing countries. The name CIVIONICS has been suggested, which connotes facilitating CIVIlians through electrONICS means. This architecture is based upon a description of technology, hardware and software (infrastructure) required for the system. In addition, the research is to leverage the abilities of available seismic sensors to develop a real-time multihazard early warning system that may be installed using realistic simulations.

Chapter 5 discusses the use of a case scenario for early warning system architecture for developing countries. The concept of Civilians Earthquake Early Warning System (CIVEEWS) has been proposed and its design parameters have been developed. The entire geographical area of the country is distributed into zones. The proposed architecture has been simulated on a computer tool.

Chapter 6 proposes the use of a case scenario for design of a multihazard early warning system for a developing country that can respond to alerts and warning messages in case of earthquake, tsunami, flood, fire, explosion, and so forth. The chapter describes the communication infrastructure and ICT for such a design, which is also simulated on a computer tool.

Chapter 7 discusses the use of information and communication technologies and wireless communication networks during multihazard disaster response phase scenarios. These technologies are Wi-Fi, WiMAX, ZIGBEE, satellite and other LTE fifth-generation networks that are useful in forming ad hoc networks and hastily formed networks (HFN) in LAN, WAN and GAN (global area network) environments. The chapter designs and develops simulated engineering scenarios and the use of these technologies in case of multihazard disaster occurrence.

Chapter 8 summarises the main contributions from this research work. It also discusses how to manage vulnerability to natural and man-made disasters through modern information and communication technology, which falls under the scope of this research work and hence tends to embark upon the goals to achieve. The natural hazards, their vulnerability or both, can cause a failure of the technology so engineers must design systems that will sustain the devastating impacts of disasters.

Finally, we welcome all comments and suggestions from readers and would love to see their feedback.

Professor Dr. Syed Hyder Abbas Musavi

Acknowledgments

First, I would like to express my deep gratitude to the Higher Education Commission Islamabad, Pakistan, which has awarded me the Indigenous Scholarship for pursuing my PhD studies at Hamdard University, Karachi, Pakistan, as most of the contents of this book are reprinted from my PhD thesis. It is also my pleasure to express my deep gratefulness to Professor Dr. Abdul Rehman Memon for his energetic and valuable guidance in the completion of my PhD degree. Without his scientific zeal, ever-willing and determined assistance, his kind solicitude for my tribulations and affable appreciations of my limitations, it would have never been viable to bring these pages to the light of day. I will remain ever grateful to him for his valuable guidance and intellectual competence. My profound appreciation to an important personality, Professor Dr Bhawani Shanker Chaudhri, ex-dean, Faculty of Electrical, Electronics and Computer Engineering, Mehran University of Engineering & Technology, Jamshoro, for his continued help in data collection, analysis and shaping of this book. In the end, it would not be justified if I do not acknowledge my friend and colleague Mr. Sikandar Ali Shah, lecturer, FEST Indus University for his valuable assistance and in co-authoring a research paper on the technical aspects of emergency telecommunication. I am also thankful to the many researchers who preceded me in this field and who paved the road upon which I travelled. Last but not the least, I am thankful to M/S OPNET Inclusion USA for providing our university with the software support of OPNET University Program Version which has been extensively used in this book. Many thanks to M/S AGI Graphics USA for their software support in the provision of STK (Satellite Tool Kit), and to Mr. Lary Cochrane of the United States who provided me WinQuake Version 3.1.4b software. I also wish to express my gratitude to Sahana Foundation Edinburg for allowing me to use the Sahana Disaster Management Tool in this work.

Author

Syed Hyder Abbas Musavi, Dr Engr, has a PhD and MS in telecommunication engineering and a BE in electronics engineering. He has been serving as Dean, Faculty of Engineering Science and Technology, Indus University, Karachi for the past four years. He has 27 years of teaching and research experience. Previously, he was engaged as chairman of the Department of Electrical and Electronics Engineering, Hamdard University, Karachi. In the past, he has served as professor and principal at Petroman, an Institute of Ministry of Information Technology and Telecommunications, Government of Pakistan at its various campuses for more than 10 years, and had also remained the executive district officer, IT. He has to his credit more than 100 research publications in national and international journals and conferences. He has attended numerous international conferences as a keynote speaker, as well as an author. He is on the review board of numerous impact factor international journals. He is a member of numerous national and international societies including IEEEP Karachi local council, IEEE USA, IEEE Computer Society, IEEE Signal Processing Society, IEEE Devices and Circuits Society, IEEE Communications Society, IEEE Education Society, to name a few.

1 Introduction

INTRODUCTION

The importance of an efficient information and communication technology (ICT) infrastructure can be realized from the frequent occurrence of natural and man-made disasters the world is witnessing at a speed not noticed before. The buoyancy of the communication infrastructure is crucial for the prosperity of any country. It is necessary to build a robust and interoperable ICT setup that will be responsible for facilitating the first responders, as well as the masses when and after the disaster hits.

Pakistan is one of the most disaster-prone countries in the world. Earthquakes, floods, tsunamis, landslides, cyclones and floods are the major threats for Pakistan. The world community was struck with horror by witnessing the devastating floods in 2010 and earthquake on October 8, 2005 in Kashmir and its adjoining areas in Pakistan. For the earthquake, it is believed that if an effective early warning system was in place in the Indian Ocean region, the impact of losses sustained by lives and property would have not been so extensive. The Government of Pakistan approved the establishment of a *National Disaster Management System* in Pakistan on February 1, 2006, which is now active in dealing with the national issues pertaining to disaster preparedness, planning, response and recovery under the umbrella of the National Disaster Management Authority (NDMA). NDMA has also strived to stretch its base to the four provinces of the country.

It was due to the will of UNESCO that on March 27, 2006, a commitment was made by the participating nations, which included Pakistan, in Bonn, Germany to establish a consortium for mitigating disaster impacts by developing national plans in lieu thereof for the requesting countries. In this meeting, the United Nations International Strategy for Disaster Reduction (ISDR) pointed out a seven-point agenda for promoting national capacities for which the consortium may provide funding. The establishment of early warning systems at the country level and disaster management short- and long-term plans is included in the said agenda. Chaudhry [1] submitted an estimate of US$38.25 million. However, the document does not speak of any details about the technology and validation of the proposal in terms of its technical success.

Early warning systems are now operational in some countries of the world. Nevertheless, the day-to-day innovations in science and technology are producing new thoughts for making the disaster management issue a problem to be solved using an engineering mind-set. The aim of this chapter, therefore, is to introduce new technological trends which will be state-of-the-art and sophisticated technological designs aimed at increased mitigation of pre- and post-disaster losses in Pakistan. The proposed system, if implemented at the national level, will give rise to a sense of security and well-being among the public. This book discusses disaster management, early warning systems, the role of information and communication technology, disaster engineering, multihazard early warning system, response measures and interoperability issues in view of any imminent threats of a large-scale disaster.

1.1 BACKGROUND

Improvements in the field of disaster engineering have been elicited following the worldwide web of natural and man-made havocs such as Hurricane Katrina, Pakistan's 2005 earthquake, numerous day-to-day bombing incidents in Pakistan and the 9/11 attacks [2,3]. The role of an efficient and robust ICT system is essential for reduction of response time and for augmenting meaningful usage of resources during the disaster management phases of relief, response, recovery and rehabilitation. There seems to be an increasing feeling among the general public worldwide for being informed ahead of any disaster to save lives and property.

Besides, the shift of the United Nations attention toward disaster mitigation measures also signifies the communications interoperability scheduling. Pakistan, along with other countries, is playing a role in the disaster-charged environment. Here, first responders are mostly active in a field with no or the least ICT infrastructure, whereby these agencies are required to equip themselves with ICT capabilities necessary for coping with challenging situations.

1.2 DISASTER MANAGEMENT

The important aspects of disaster management engineering are to develop a generic design of a multihazard early warning system, and to model architecture for an integrated interoperable wireless communication network for disaster responders (IWCNDR), in addition to the use of information and communication technology in disaster management operations.

1.3 ECONOMIC BURDEN OF DISASTERS

Between 1998 and 2017, due to a geophysical disaster, earthquakes, tsunamis and floods killed about 1.3 million people and left more than 4.4 billion injured, homeless, displaced or in need of emergency aid.

The World Bank has calculated that the real cost to the global economy is a staggering US$520 billion per annum, with disasters pushing 26 million people into poverty every year. While high-income countries reported losses of 53% of disasters between 1998 and 2017, low-income countries only reported 13% from disasters. No loss data are therefore available for nearly 87% of disasters in low-income countries [7]. At present, when climate change is increasing the frequency and severity of intense weather events, disasters will continue to be the main impediments to sustainable development, so long as the economic incentives to develop in hazard-prone locations such as floodplains, vulnerable coasts and earthquake zones continue to outweigh the perceived disaster risks.

Globally, geophysical disasters—primarily earthquakes—killed more people than any other type of natural hazard in the past 20 years, with a cumulative toll of 748,000 fatalities. This was exacerbated by the vulnerability of poor and badly prepared populations exposed to two major events: the 2004 Indian Ocean tsunami and the Haiti earthquake of 2010. Floods also affected the largest number of people, at more than 2 billion, followed by drought, which affected another 1.5 billion people between 1998 and 2017 [7].

In a review undertaken in 53 developing countries over the period 2006–2016, the FAO found that, taken together, crops and livestock, farming, fisheries, aquaculture and forestry absorbed 26% of all damage and loss caused by floods, drought and tropical storms. Almost two-thirds of all damage and loss to crops was caused by floods in that decade, but drought was by far the most harmful disaster for livestock. The year 2011 reflects the immense damage caused by the Great East Japan Earthquake and Tsunami, with the consequent shutdown of the Fukushima nuclear energy plant, with losses totalling US$228 billion. In 2008, the earthquake in Sichuan, China, cost US$96 billion and affected 46 million people [7]. Population growth, economic development and rapid and often risk-blind urbanization place more people in harm's way than ever before in earthquake zones, floodplains, coastlines, dry lands and other high-risk areas, increasing the possibility that a natural hazard will turn into a humanitarian catastrophe. More people are affected by extreme weather events, especially floods and drought, than by any other type of natural hazard. While early warning systems and timely evacuations have led to reduced loss of life, economic losses continue to grow, impeding a number of nations graduation from least-developed country (LDC) status to middle-income status.

1.4 MODERN INFORMATION TECHNOLOGY USAGE TO MANAGE DISASTERS

Engineering systems, technologies, methods, techniques, inventions and innovations are believed to be crucial to mitigate disaster losses [2–4]. Hence, the role of technology in disaster management is unequivocal and inevitable. This text, therefore, examines these functions by taking into account various factors vulnerable to disaster losses. Apart from economic, psychological and educational tools, technology can contribute, suggest and improve the state of affairs of natural disasters when and where these hit throughout their life cycle. Public consciousness through mass media is performed through technology and the examples of many American films on disasters can be thought of as a helping point toward disaster management. Engineers create and use technology through research, development, design and analysis. During the pre-disaster phase (mitigation, prevention, preparation and planning), the examples of technology are *monitoring, prediction and early warning systems* along with *information and education* about technology and natural disasters, whereas during the post-disaster phase (response, recovery and rehabilitation), the technology examples are telemedicine and health facilities, equipment and devices related to rescue and response, telecommunication, mapping, GIS, *satellite, wireless, hastily formed networks*, email, Internet, and so forth.

1.5 EARLY WARNING SYSTEMS

A UN Survey Report on global early warning systems published on March 26, 2006 has voiced concerns about the shortcomings in warning systems around the world's developing countries. This survey suggests global early warning in concurrence with the existing ones possessing advanced technology attributes. This report focuses on capabilities of early warning systems with regard to warning messages which report, suggesting that those should be reachable to all concerned. The report points out the following root causes in the delay in development of early warning systems and their messages [5,6]:

- Public unawareness to realize vulnerabilities to their disasters
- Lack of political will
- Weak coordination among various actors

The above statement is highly valid for developing nations like Pakistan. We have witnessed that people did not receive any warnings in the case of the October 8 Kashmir earthquake, and the same is true when the people of Khyber Pakhtunkhwa did not receive warnings or were unable to understand the severity of its messages and consequences during the devastating floods in 2010. Furthermore, the authorities concerned should have thought upon the concept of a multihazard early warning system, rather than one acting and dealing with disasters individually and not focusing a coherent integrated strategy to deal with these disasters. Therefore, there arises a need to clearly define such early warning systems which pose timely, successful and meaningful warning signals for all types of disasters to all at risk and to the first responders. In our work, we have tried our best to address this issue by discussing the design of such types of early warning systems.

SUMMARY

This chapter discusses the need for an early warning system for multihazard scenarios. The chapter also covers the economic burden of disasters on a global level. The behavior of people not responding to disaster literacy measures contributes to the loss of human lives and burdens to the global economy in billions. Hence, the need for initiating training related to the disaster mitigation phase is also discussed. The author emphasizes the need for more initiatives to be taken for establishing ICT-based multihazard early warning systems in developing countries such as Pakistan, Afghanistan, Sri Lanka, Iran and India.

REFERENCES

1. Chaudhry, Q., "Strengthening National Capacities for Multi Hazard Early Warning and Response System Phase I," Cabinet Division Government of Pakistan, Islamabad, 2006. Available at http://www.pmd.gov.pk/Establishment%20of%20Early%20Warning%20System.pdf
2. Dhaini, A.R., Ho, P.-H., "MC-FiWiBAN: An Emergency-Aware Mission-Critical Fiber-Wireless Broadband Access Network," *IEEE Communications Magazine*, January 2011: pp. 134–142.
3. Chris Oberg, J., Whitt, A.G., Mills, R.M., "Disasters Will Happen—Are You Ready?" *IEEE Communications Magazine*, January 2011: pp. 36–42.

4. Mase, K., "How to Deliver Your Message from/to a Disaster Area?" *IEEE Communications Magazine*, January 2011: pp. 52–57.
5. Anan, K.A., "Global Survey of Early Warning Systems," United Nations, 2006. Available at http://www.unisdr.org/ppew/info-resources/ewc3/Global-Survey-of-Early-Warning-Systems.pdf
6. United Nations Economic and Social Council, "Enhancing Regional Cooperation on Disaster Risk Reduction in Asia and the Pacific: Information, Communications and Space Technologies for Disaster Risk Reduction," March 25–27, 2009.
7. Economic Losses, "Poverty and Disasters 1998–2017 Report by (UNISDR) and (CRED)," Available at https://www.unisdr.org/we/inform/publications/61119

2 Disaster Management and Early Warning Systems

INTRODUCTION

The terms Disaster and Hazard are usually confused. Disasters are infrequent, occurring in a society that is vulnerable to Hazards like floods, hurricanes, earthquakes, and so forth. Hazards are technological, wilful (like blasts) or man-made and natural. The occurrence of a disaster is termed an Incident. Risk is the by-product of an Incident and Hazard [1–3].

Disaster Management involves four phases: preparedness, planning, response and recovery. In the disaster cycle, the steps involved are Mitigation (Prevention, Preparedness and Planning), Alertness, Response and Recovery (Rehabilitation and Reconstruction). Disaster Management refers to a collective term encompassing all aspects of planning for and responding to disasters including both the pre- and post-disaster activities. It refers to both the risk and consequences of disaster [1]. According to the World Disasters Report (published by "International Federation of Red Cross in 2005" [23]), forecasting, national warning, local government diffusion, civil society participation and popular understanding and action need to come together for successful disaster management. As stated in this report, "nowhere, can the government do everything." Civil society must hurl in. But this means people must trust the government. Furthermore, the disasters generate heavy economic and human losses. For every region on earth, studies are necessarily to be undertaken to expose the vulnerability of that region to particular disaster(s) due to certain risks. Disasters are the functions and matrices of risks and vulnerabilities. The greatest potential for maximum reduction of losses is during the Mitigation Phase. All tools, techniques, methodologies, systems and approaches should be focused in this phase. Next is the response phase, which is also crucial and should be managed with top expertise failing which losses more than those occurred in the mitigation phase may be faced. Examples include not shutting down the operation of trains after receiving an earthquake warning signal and not closing the valves of gas stations during the same disaster or not winding up power supply to an atomic or critical infrastructure. All these actions require careful planning and design of efficient systems with trained manpower responsible to operate them.

2.1 DISASTER MANAGEMENT AND DISASTER RESILIENCE

During the mitigation phase, it is widely recommended to plan and prepare the nation for natural disasters, its impacts and consequences [4,5]. This is also referred to as "Preventive Engineering" [6]. This involves handling the issue at its root cause. The root cause of natural disasters depends on Risk, which in consequence is the function of hazards and vulnerability. Therefore, preventive engineering encompasses two major factors: resilience to natural hazards and resilience to vulnerability [6]. The discussion regarding these two factors is out of the scope of this work; however, one should understand the process of evolution, conduct and results of natural hazards. In order to resist too much damage from natural disasters, society should manage vulnerability by applying and adopting appropriate technology. Examples of actions for tackling the mitigation phase through technology are:

i. Designing and implementing building codes for earthquake disasters and wind storms
ii. Land-use planning and zoning, its legislation to prevent people from seeking settlement alongside the flood-prone areas
iii. Designing information and communication networks for efficient delivery of disaster reporting and coordination
iv. Designing the monitoring and early warning systems
v. Education and Training

Tasks (iii) and (iv) from the above-mentioned example do come within the scope of this work.

2.2 REVIEW OF EARLY WARNING SYSTEMS

The important international efforts that have been made with a focus on early warning systems are:

- The three global early warning conferences (1998, 2003 and 2006)
- The World Conference on Disaster Reduction, held in Kobe, Hyogo, Japan, during January 18–22, 2005 adopted the Hyogo Framework for Action 2005–2015: building the resilience of nations and communities to disasters (published in "World Disaster Report 2009" [24])
- In 2006, the World Meteorological Organization (WMO) conducted an "Assessment of National Meteorological and Hydrological Services" capacities in support of disaster risk reduction [7]
- Functioning of World Meteorological Organization (WMO) [7] with its data collection and sharing networks including the Global Observing System, Global Telecommunications System [8] and global data processing and forecasting system, and the UN International Strategy for Disaster Reduction (UNISDR) [7] which promotes policy, strategic and programmatic work on disaster risk reduction
- Notable single hazard systems include the Japanese Government and WMO's efforts to monitor floods globally, with the Global Flood Alert System, under the International Flood Network. The UN's Food and Agriculture Organization leads efforts to track food insecurity through the

Global Information and Early Warning System, whereas the World Health Organization (WHO) leads global mechanisms to issue health-sector early warnings and crucial coordination efforts are under way to improve tsunami warnings for various oceans, under the Intergovernmental Coordination Group operating through the United Nations Educational, Scientific, and Cultural Organization (UNESCO) [8]. The Global Tropical Cyclone Warning System is one of the best examples of international, regional and national collaboration in technical monitoring and warning. The WMO's global operational network enables continuous observation, data exchange and regional forecasting [8]

- There are six centres responsible for forecasts of meteorological data, warning alerts and news to national meteorological services in all countries at risk with lead times of 24–72 hours. The national services then issue warnings to governments, the media and the general public according to national protocols (published in "World Disaster Report 2009" [24])
- These international efforts have contributed a lot of awareness among the public and governmental bodies of developing nations to sense the disaster issue with a technological mind-set and thus have led to increased interest of these agencies toward managing the disaster engineering issue efficiently

2.3 EARLY WARNING SYSTEMS—THE INTERNATIONAL SCENARIO

This section focuses on the existing seismic early warning systems in place within advanced countries and in a few developing nations.

Two different approaches are in operation worldwide for implementing an earthquake early warning system. The first approach [25] uses a single station approach where seismic signals are processed locally and an earthquake warning is issued when the ground motion exceeds the triggered threshold. The second approach [26] is called the array approach where a central station processes signals from an array of stations and decides whether the earthquake has occurred or not [9].

Thanks are due to J.D. Cooper who in 1868, introduced for the first time the concept of an early warning system for San Francisco, California. The modern seismic alert networks, which are mostly computerized, were proposed by Heaton in 1868. Currently, Earthquake early warning systems are either implemented or under construction or in the planning stage in Mexico, Romania, California, Japan, Taiwan, Italy, Turkey and Greece [10,11]. Such systems are still very much in progress: Japan, Mexico and Taiwan are deploying them, but most other nations, including the United States, are still in the research stages, according to leading seismologist Haroo Kanamori [11].

Earlier, such systems were called "earthquake notification systems." In the 1960s, the US Geological Survey (USGS) in Menlo Park developed an earthquake monitoring system with features to find the location and magnitude estimation of earthquakes [10]. Later, the USGS in collaboration with the California Institute of Technology developed a seismic network in southern California. Other examples from those years are the Rapid Earthquake Data Integration Project (REDI) [12] in northern California and Caltech/USGS Broadcast Earthquake (CUBE) in southern California [13–15]. With the previous technology, only messages of earthquake location and magnitude were communicated to users within minutes of occurrence of the event, but in 2001, 2002 and 2003, in southern California, a new seismic network called TriNet [16] was developed with the capability to additionally broadcast shake maps within a few minutes of the occurrence of the event. Earthquake warnings are transmitted using ICT in Japan, which owns the world's most sophisticated Earthquake Early Warning System.

Nakamura in 1988 [25] presented the concept of slowing down or stopping high-speed railway trains after receiving an early warning system from the UrEDAS system in Japan. Bakun et al. in 1994 [27], after the California 6.9 magnitude earthquake, developed an early warning system in Oakland, which gave a warning time of 20 seconds at about 100 km from the epicenter, for workers to evacuate.

Since 1985, Mexico City also possesses the advanced Seismic Alert System between Mexico City and the Guerrero seismic fault line. This system has 12 seismic stations with a spacing of 25 km. This system was designed and implemented by the Centro de Instrumentacion y Registro Sismico (CIRES) under the instructions of J.M. Espinosa-Aranda. It is composed of four segments: seismic detection, telecommunications, central control and radio warning. Each field station monitors the seismic activity within a 100 km radius and detects and estimates the magnitude of an earthquake within 10 seconds of its initiation. If the estimated magnitude is greater than 6, a warning message is sent on a telecommunications unit to the central control unit in Mexico City. A public alert signal is sent through the radio warning unit if two or more field stations confirm the occurrence of the earthquake [15,19]. The earthquake detection system in Mexico City can issue alerts regarding high magnitude seismic disasters.

Wu et al. in 2013 [28] presented the concept of a virtual subnetwork and implemented their ideas on Taiwan EWS which gave empirical results of the warning time of 20 seconds at a distance of 145 km from the epicenter.

In 2007, Weber et al. [29] proposed and implemented a dense network of seismometers and accelerometers for the Campania region of southern Italy. The project is called the Irpinia Seismic Network (ISNet) along the Apennine Belt covering an area of 100 km × 70 km, and is a regional-type system.

2.4 WORLD EARLY WARNING SYSTEMS

Figures 2.1 through 2.7 [20] show example scenarios of a few early warning systems and their concepts in rouge in developed countries like Japan. The aim of exploring these systems is to invoke ideas for developing early warning systems in Pakistan.

These charts and the discussion illustrate in detail the following key elements of an EWS:

a. Sensors to sense the impending disaster
b. Communication to base station(s) for severity assessment/action
c. Action based on severity

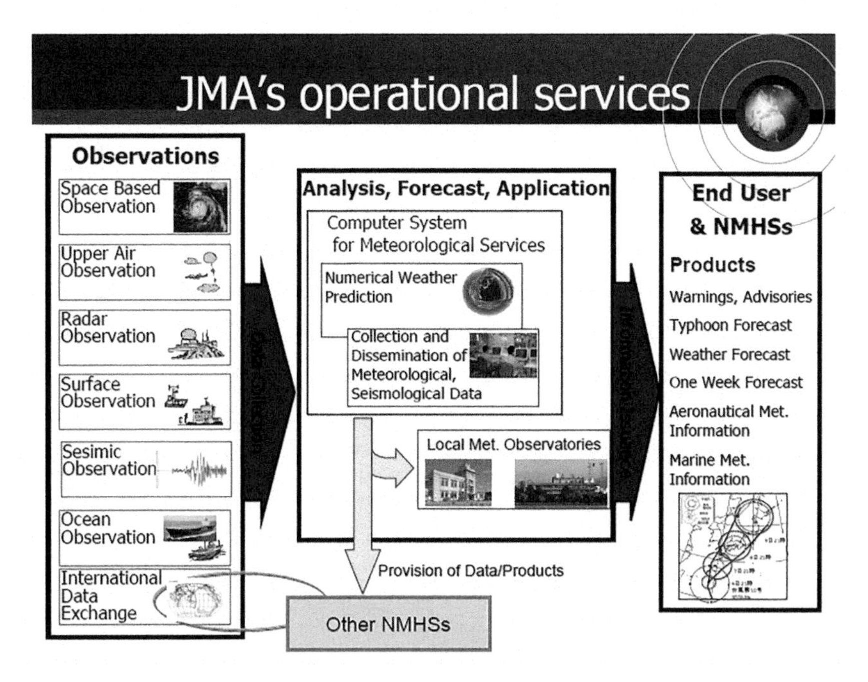

FIGURE 2.1 Meteorological operational service in Japan. (International Centre for Water Hazard and Risk Management. Japan Meteorological Agency [20].)

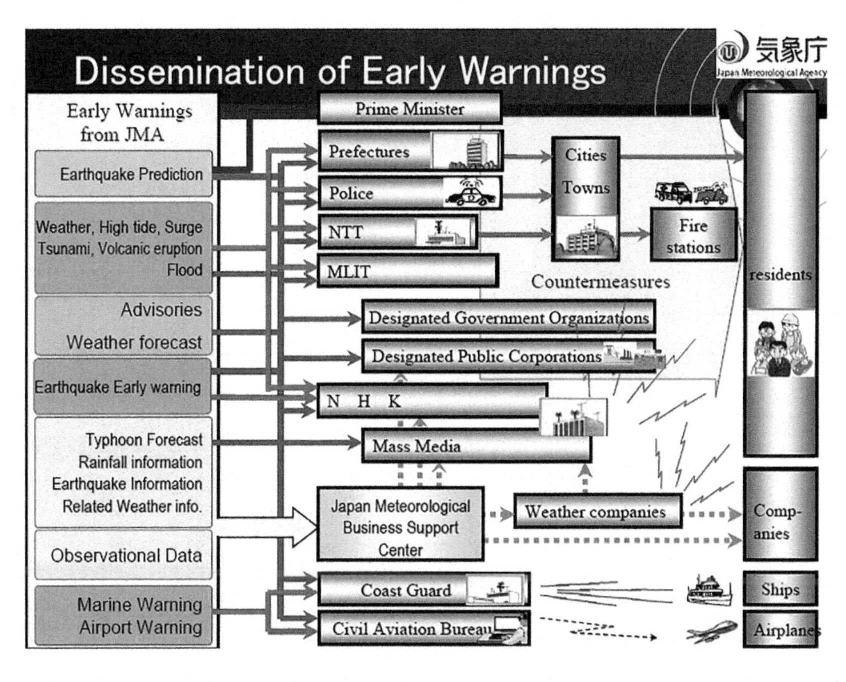

FIGURE 2.2 Disseminations of early warnings concept in Japan. (International Centre for Water Hazard and Risk Management. Japan Meteorological Agency [20].)

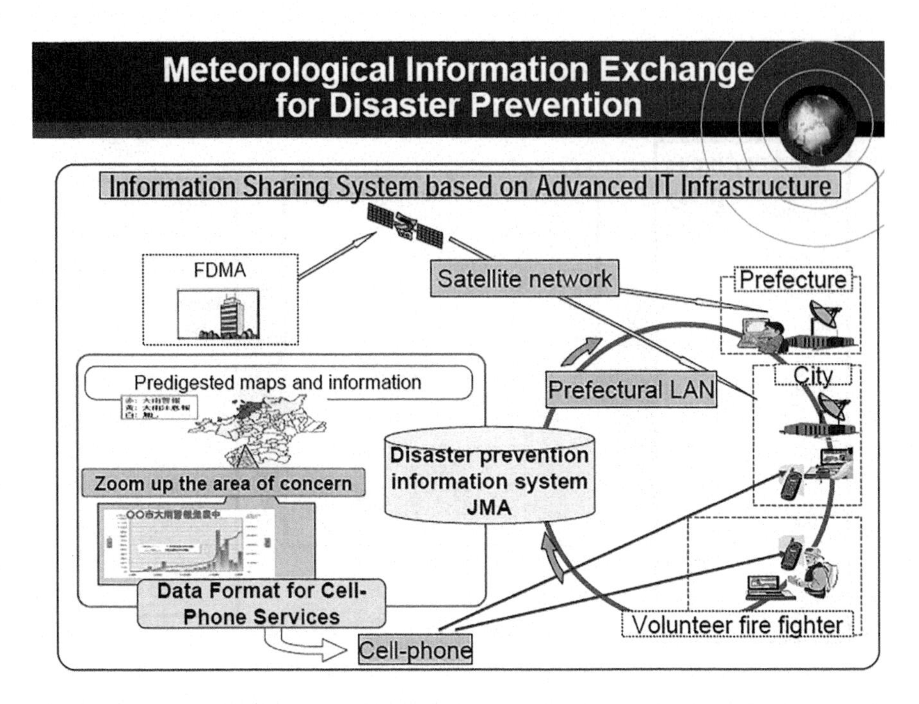

FIGURE 2.3 IT infrastructure for information sharing in Japan. (International Centre for Water Hazard and Risk Management. Japan Meteorological Agency [20].)

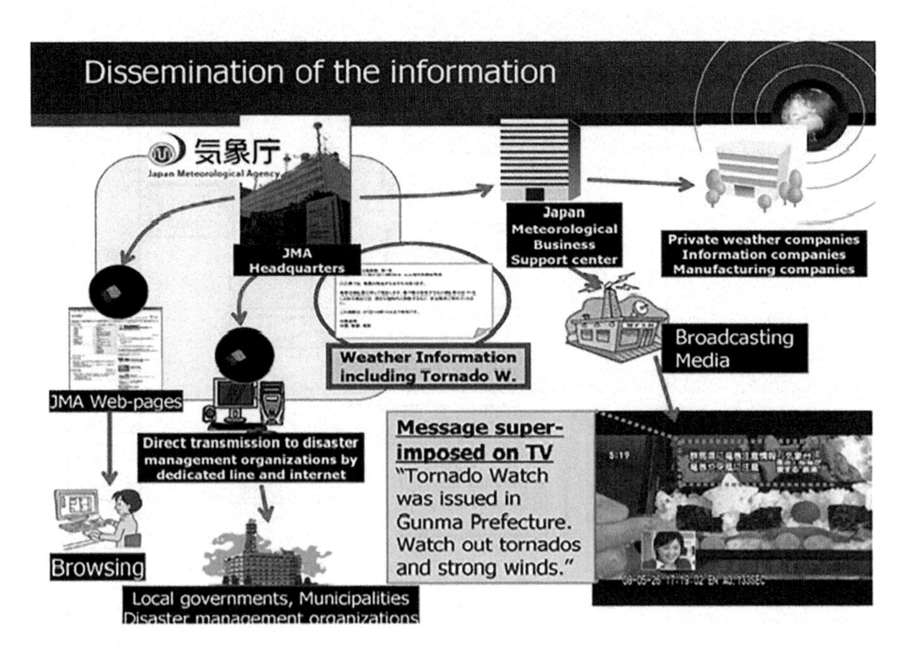

FIGURE 2.4 Example of information dissemination in Japan. (International Centre for Water Hazard and Risk Management. Japan Meteorological Agency [20].)

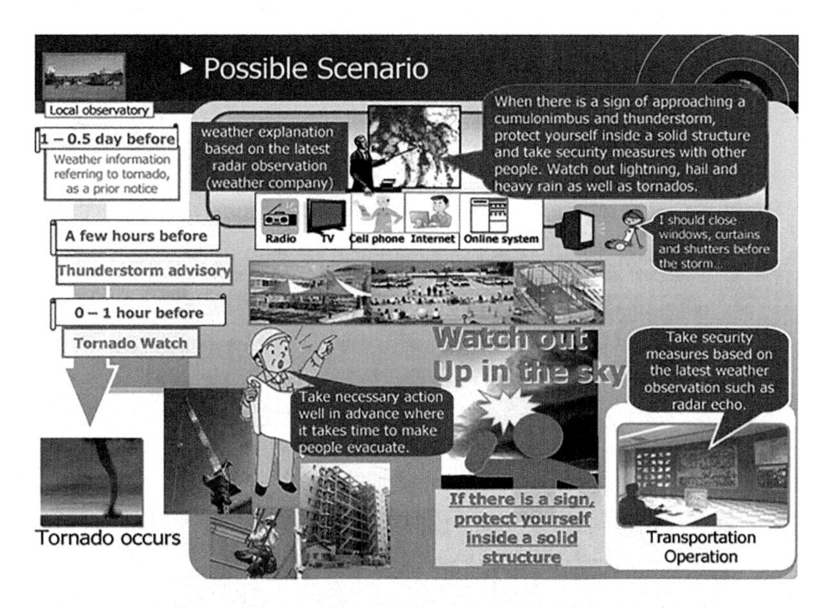

FIGURE 2.5 Early warning possible actions in disasters. (International Centre for Water Hazard and Risk Management. Japan Meteorological Agency [20].)

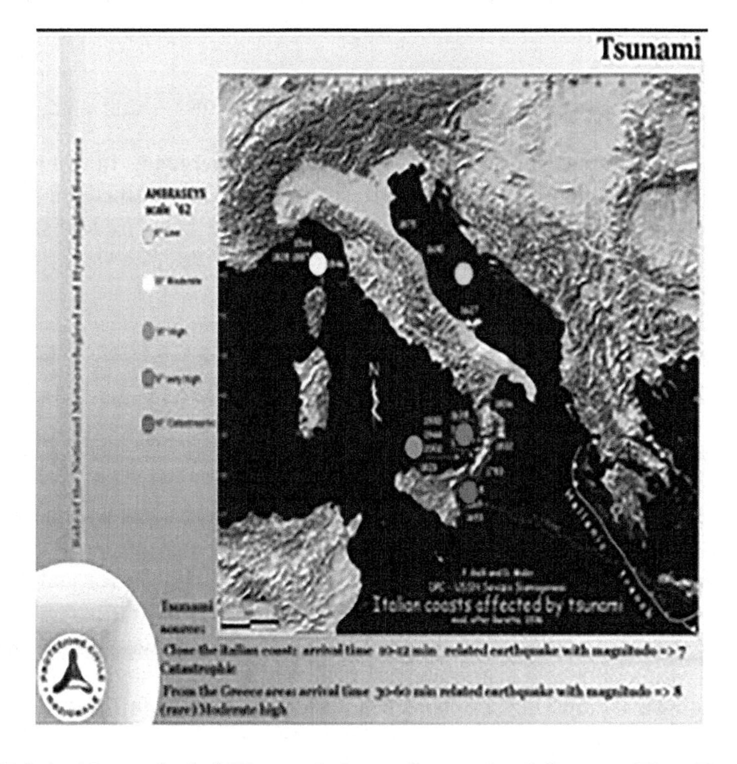

FIGURE 2.6 Meteorological 3D maps in Japan. (International Centre or Water Hazard and Risk Management. Japan Meteorological Agency [20].)

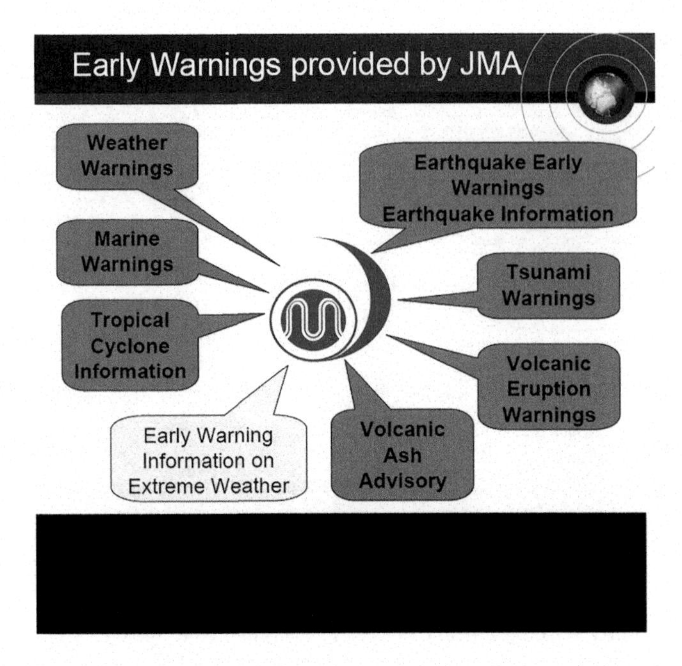

FIGURE 2.7 Types of early warnings by JMA. (International Centre for Water Hazard and Risk Management. Japan Meteorological Agency [20].)

2.5 DISASTER RESPONSE AND IT USAGE

The Disaster Resource Network in a report titled "Emergency Information and Communications Technology in Disaster Response" [2] has identified the following key factors that impact the effectiveness of emergency ICT response to disasters:

i. Equipment and technology requirements
ii. Inter- and intra-agency cooperation
iii. Personnel
iv. Standards and governance
v. Interoperability
vi. Sustained advance preparedness

The integrated community-based disaster management concept was put forward [1] according to which the response activity results can be improved using this idea. The constant development in the potential of mobile technologies and wireless communications has considerably changed how the mobile and dispersed work of emergency response could be maintained [3,21]. In modern days, when network connectivity for mobile and distributed actors is easy, the focus has been transferred to investigating and preparing design doctrines and provisions on information systems for emergency response management [3]. With the aim to develop integration and interoperability between mobile and centralized information systems and to improve coordination among all disaster responders, the thrust for a secure and redundant ICT setup is increasingly realized [3].

The use of redundant emergency communication systems has been emphasized in response activities [8,21]. These systems include terrestrial fixed services, satellite mobile phones, mobile and wireless services, ham radio (amateur services), televisions, radios, SMS, public warning and notification systems, GIS, earth observatory satellite services, GSM, VoIP, software like the Sahana Disaster Management system, as well as Internet emergency communication vehicles connected with satellites, remote sensing, videoconferencing and telemedicine devices, equipment, software and hardware.

Interoperability of devices, communication systems and networks is the key to successful response activities. This can be achieved through imparting a viable and efficient thought given to a nationwide interoperable communication network dedicated to response activities. Examples of such networks include the Federal Communications Commission's (FCC) 911 calls, wireless 911 calls, VoIP services, Emergency Alert System (EAS) and Commercial Mobile Alert System (CMAS) [14,22].

2.6 THE INTEROPERABILITY ISSUE

In general, interoperability refers to the ability of emergency responders to work seamlessly with other systems or products without any special effort [18]. Wireless communications interoperability specifically refers to the ability of emergency response officials to share information via voice and data signals on demand, in real time, when needed, and as authorized [18]. In such systems, the first responders can talk across the board in a coordinated manner permitting them to share their resources cooperatively during the disaster rescue, relief and recovery phase.

2.7 CHALLENGES TO INTEROPERABILITY

The following are considered to be the fundamental problems as far as disaster engineering communications technologies usages are concerned [18]:

- Mismatched and old devices
- Shortages in budgets, finances and funding
- Issues in planning and coordination
- Non-allocation of appropriate bandwidth and radio spectrum
- Equipment standardizations

To comply with the interoperable communications network requirement, a model is required that should be a common architecture between all the agencies operating in Pakistan in disaster management. The discussions shall also include the interface components of such a system.

2.8 INTERAGENCY COMMUNICATION PROBLEMS

In the United States, the Office for Interoperability and Compatibility (OIC) of the Department of Homeland Security (DHS) started a national communications program (SAFECOM) to focus on interoperability problems. The National Task Force on Interoperability (NTFI), comprising 18 national associations, bears the mandate to address issues pertaining to communications interoperability. The SAFECOM program has pointed out the following important elements related to inter-agency communications problems:

- Technology
- Frequency of Use
- SOPs (Standard Operating Procedures)
- Governance
- Training and Exercises

In addition, the following technical means have been suggested by the SAFECOM program for inter-agency communications:

- Sharing Standard-Based Systems
- Sharing Channels
- Sharing Proprietary Systems
- Swap Radios
- Using Gateways between Independent Systems

Furthermore, according to SAFECOM [18,21], the scenarios where the agencies cannot talk to each other may fall into the following categories:

- Swap Radios (SR)
- Talkaround (TKR)
- Mutual Aid Channel (MAC)
- Gateway Console Patch (GCP)
- Network Roaming (NR)
- Standards Based Shared Networks (SBSN)

In the following chapters, the above-mentioned concepts will be explained along with sections that provide answers to questions like: What is needed for security agencies to network? What should be the core attributes of such a network; coverage, accessibility, transmission speed, prompt call setup, network capacity, mobility, and so forth? What ICT services are required for such an interoperable system in Pakistan?

SUMMARY

This chapter discusses the phases of disaster management and information and communication technology in early warning systems. It elaborates the challenges being faced by an intra-agency communication network during the response phase, and what ICT services are required for such an interoperable system in developing countries. A discussion about the world's efficient early warning systems is also provided.

REFERENCES

1. Østensvig, I., "Interagency Cooperation in Disaster Management," in *Department of International Environment and Development Studies*, Norwegian University of Life Sciences, Jamaica, 2006. Available at www.jamaicasugar.org/.../Section%206%20-%20 Disaster%20&%20Emergency%20Response.pdf
2. Disaster Resource Network, "Emergency Information and Communications Technology in Disaster Response, Final Report," 2007. Available at www.drnglobal.org/news/ drn-releases-report-on-emergency-ict-in-disaster-response/
3. Landgren, J., "Designing Information Technology for Emergency Response," *A Doctoral Dissertation of Department of Applied Information Technology*, IT-University of Goteborg, Goteborg University: Goteborg, Sweden, 2007. Available at www.ituniv.se; www.gu.se; www.chalmers.se
4. Dhaini, A.R., Ho, P.-H., "MC-FiWiBAN: An Emergency-Aware Mission-Critical Fiber-Wireless Broadband Access Network," *IEEE Communications Magazine*, January 2011: pp. 134–142.
5. Krock, R.E., "Lack of Emergency Recovery Planning-Is a Disaster Waiting to Happen?" *IEEE Communications Magazine*, January 2011: pp. 48–51.
6. Kelman, I. "Role of Technology in Managing Vulnerability to Natural Disasters," *A Masters of Applied Science thesis*, Graduate Department of Civil Engineering, University of Toronto, 1998.
7. United Nations Economic and Social Council, "Enhancing Regional Cooperation on Disaster Risk Reduction in Asia and the Pacific: Information, Communications and Space Technologies for Disaster Risk Reduction," March 25–27, 2009.
8. Anan, K.A., "Global Survey of Early Warning Systems," United Nations, 2006. Available at http://www.unisdr.org/ppew/info-resources/ewc3/Global-Survey-of-Early-Warning-Systems.pdf
9. Lee, W.H.K., *"Earthquake Early-Warning Systems: Current Status and Perspectives,"* United States Geological Survey, Menlo Park, CA 94025, USA. Available at http://www.cires.org.mx/docs_info/CIRES_007.pdf
10. Aldo Zollo, D.C.S., "Real Time Location for a Seismic Alert Management System-Development, HW/SW Integration, Definition and Study of Velocity Models," PhD Thesis, Universit`a di Bologna: Campania Region, Southern Italy, 2008.
11. MIT Technology Review. Available at http://www.technologyreview.com/infotech/20772/?a=f [Last accessed on December 15, 2010].
12. Rapid Earthquake Data Integration Project (REDI) [7] in northern California. Available at http://seismo.berkeley.edu/redi/redi.overview.html [Last accessed on December 15, 2010].
13. Hiroo Kanamori, E.H.T.H., "Real-Time Seismology and Earthquake Hazard Mitigation," *Nature*, Vol. 390, 1997. Available at http://ecf.caltech.edu/~heaton/papers/kanamori_realtime_nature.pdf
14. Kazakhstan, A. "*Report of the Gcos Regional Workshop For Central Asia on Improving, Observing Systems for Climate*," Available at http://www.wmo.ch/pages/prog/gcos/Publications/gcos-94eng.pdf [Last accessed on December 15, 2010].
15. Erdik, M., "Urban Earthquake Rapid Response and Early Warning Systems," *First European Conference on Earthquake Engineering and Seismology*. September 3–8, 2006: Geneva, Switzerland.

16. Seismic Project. Available at http://www.trinet.org/ [Last accessed on December 15, 2010].
17. Nakamura, Y., "Uredas, Urgent Earthquake Detection and Alarm System, Now and Future," *13th World Conference on Earthquake Engineering*, Vancouver, B.C., Canada, 2004. Available at http://www.citeseerx.ist.psu.edu/viewdoc/download?doi=10.1.1.165.5764
18. SAFECOM, Available at http://www.safecomprogram.gov/SAFECOM/interoperability/default.htm
19. Seismic Laboratory Online Documentation. Available at http://www.seismolab.caltech.edu/early/Real-Time_Seismology_and_Earthquake_Damage_mitigation4.pdf, [Last accessed on December 15, 2010].
20. JMA, "Overview of Early Warning Systems and the Role of National Meteorological and Hydrological Services Japan," *International Centre or Water Hazard and Risk Management (ICHARM)*. Available at www.wmo.int/pages/prog/drr/...II/.../01-MHEWS-II-Japan.pdf
21. Chris Oberg, J., Whitt, A.G., Mills, R.M., "Disasters Will Happen—Are You Ready?" *IEEE Communications Magazine*, January 2011: pp. 36–42.
22. Federal Communications Commission, USA Emergency Communication Website. Available at http://www.fcc.gov/cgb/consumerfacts/emergencies.html [Last accessed on December 15, 2010].
23. International Federation of Red Cross. 2005. Available at https://www.unisdr.org/2006/ppew/info-resources/ewc3/Global-Survey-of-Early-Warning-Systems.pdf.
24. World Disaster Report. 2009. Available at http://www.unisdr.org/2006/ppew/info-resources/ewc3/Global-Survey-of-Early-Warning-Systems.pdf.
25. Nakamura, Y., "On the Urgent Earthquake Detection and Alarm System (UrEDAS)." *Proceedings of the 9th World Conference on Earthquake Engineering*, 1988. Tokyo-Kyoto, Japan.
26. Lee, W.H.K., Shin, T.C., Teng, T.L., "Design and Implementation of Earthquake Early Warning Systems in Taiwan." *Proceedings of the 11th World Conference on Earthquake Engineering*, 1996. Acapulco, Mexico.
27. Bakun, W.H., Fischer, F.G., Jensen, E.G., VanSchaack, J., "Early Warning System for Aftershocks." *Bulletin of the Seismological Society of America*, 1994: 84(2).
28. Wu, Y.M. et al., "Earthquake Early Warning System in Taiwan." In: Beer, M. et al. (eds.) *Encyclopedia of Earthquake Engineering*, 2013: pp. 1–12. Berlin: Springer.
29. Weber, E. et al., "An Advanced Seismic Network in the Southern Apennines (Italy) for Seismicity Investigations and Experimentation with Early Earthquake Warning." *Seismological Research Letters*, 2007: 78(6).

3 Disaster Engineering Computer Tools

INTRODUCTION

Two important tools are usually implemented worldwide in disaster engineering. One is the Hazus Loss Estimation Tool used by the Federal Emergency Management Authority (FEMA) in the United States, and the second is the Sahana Disaster Management Tool. Both these tools are used in the response phase of disaster engineering. We recommend that these tools also be implemented in Pakistan. The Hazus loss estimation tool presently is technically valid for the geography of the United States; however, the development of a Pakistan-specific Hazus-like loss estimation tool to be viable for Pakistan's territorial jurisdiction is under progress. This book will use the Pakistan Sahana Disaster Management Tool, which is described in this section. First, however, we will discuss the salient features of both these tools.

3.1 SALIENT FEATURES OF THE HAZUS-MH LOSS ESTIMATION TOOL

- Hazus-MH is a cutting-edge software model at FEMA for estimating losses that may occur if disasters (floods, earthquakes and hurricanes) occur [1]
- Hazus-MH allows users to run what-if scenarios
- Results from Hazus-MH provide decision makers [1] with necessary information to:
    i. Assess the level of readiness and preparedness to deal with a disaster in a given region within the United States (before the disaster hits)
    ii. Decide on how to allocate resources for the most effective and efficient response and recovery when a disaster hits a certain region
    iii. Prioritize the mitigation measures that need to be implemented to reduce future losses
- Hazus-MH is still evolving to include additional hazards: airborne toxic releases, dam failures, etc.
- For more information on Hazus, please visit: www.fema.gov/hazus

3.2 SCREENSHOTS OF THE HAZUS-MH SOFTWARE

Figures 3.1 through 3.9 show a few screenshots of the Hazus-MH software used for the Disaster Loss Estimation Tool by FEMA in the United States: Hazus is used for mitigation and recovery as well as preparedness and response. Government planners, GIS specialists and emergency managers use Hazus to determine losses and the

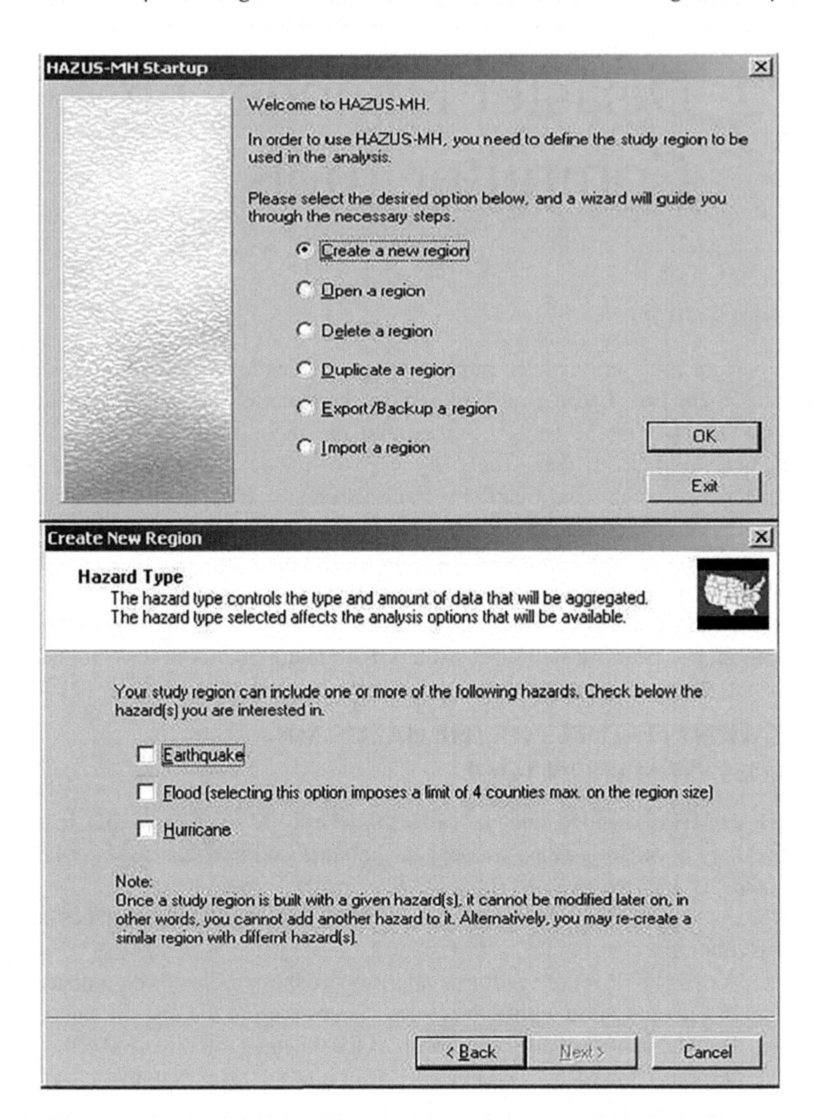

FIGURE 3.1 Step 1(a): Creating a new region (based on the Hazus-MH software).

most beneficial mitigation approaches to take to minimize them. Hazus can be used in assessment steps in the mitigation planning process which is the foundation for a community's long-term strategy to reduce disaster losses and break the cycle of the disaster damage, reconstruction and repeated damage [1].

Figures 3.1 and 3.2 show step 1, creating a new region; Figures 3.3 and 3.4 show step 2, accessing a created region; step 3 is the Hazus-MH interface shown in Figure 3.5; the Hazus-MH inventory menu is shown in Figure 3.6; the Hazus-MH analysis menu (analytical engine) is shown in Figure 3.7. Figure 3.8 presents the Hazus-MH results menu and Figure 3.9 shows the Aloha Plume Overlaying Hazus-HM Inventory.

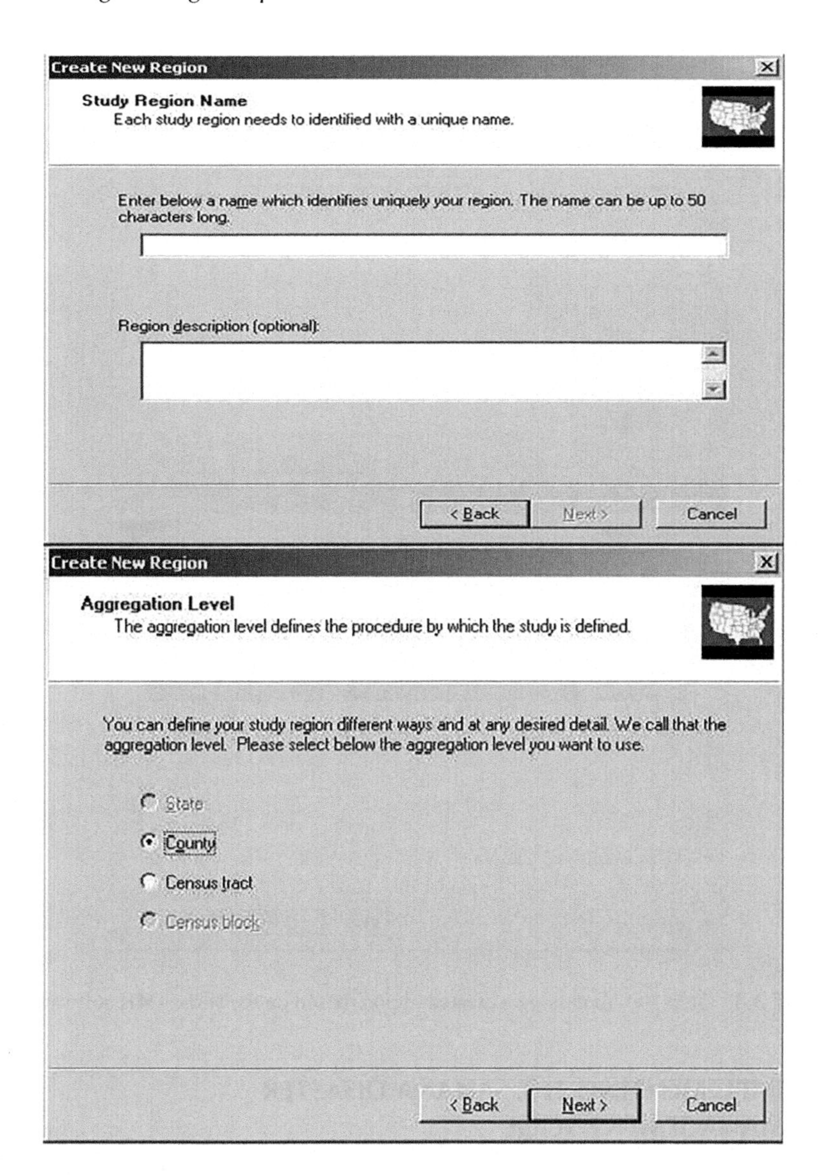

FIGURE 3.2 Step 1(b): Creating a new region (continue). (based on the Hazus-MH software).

The screenshots in Figure 3.9 pertain to the Hazus-MH Loss Estimation Tool used by FEMA in the United States. Due to its extreme need in the post-disaster scenario, our aim of producing the screenshots in this work is to apprise the authorities responsible for disaster management in Pakistan about the necessity and importance of using this ICT tool. The functions and scope of this tool were described in Section 3.1. However, as stated earlier, a country-specific loss estimation tool has to be designed by every country in the world that meets the requirements of their own territorial boundaries.

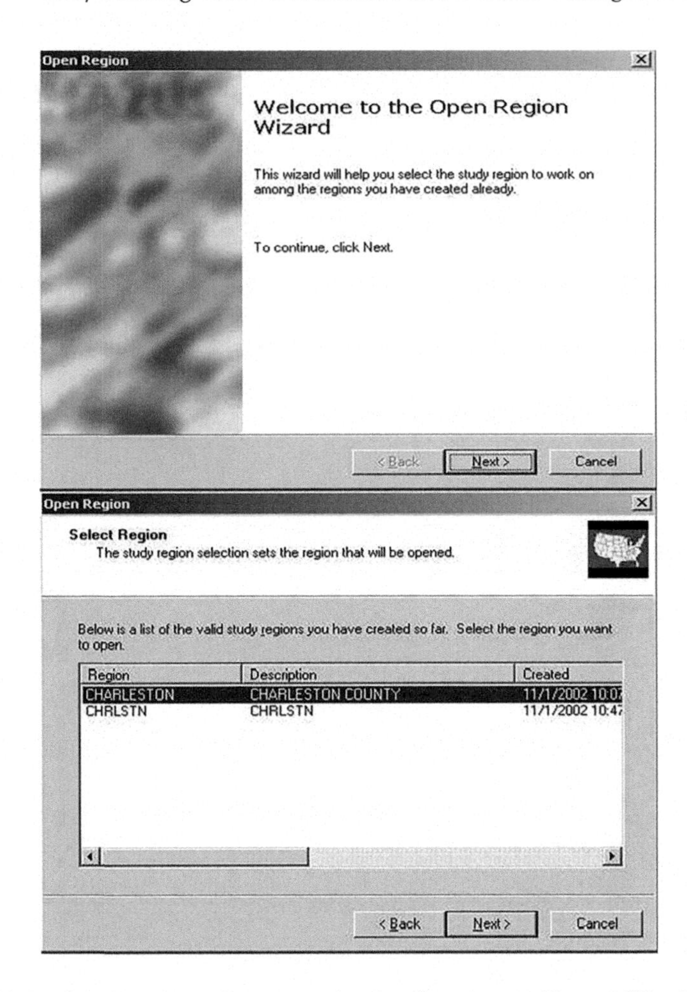

FIGURE 3.3 Step 2(a): Accessing a created region (based on the Hazus-MH software).

3.3 IMPLEMENTING THE SAHANA DISASTER MANAGEMENT TOOL

Due to the courtesy of the Sahana Foundation Edinburg, the author was provided access to the computer-based Sahana Disaster Management Tool. We have used this tool in research and have found that it is beneficial in disaster engineering before, during and after an emergency hits any country.

3.3.1 Application of Sahana for Disaster in Pakistan

In the aftermath of the floods of 2010 in Pakistan, the author used the Sahana Disaster Management Tool screenshots are depicted in Figures 3.10 through 3.14. A brief explanation of the work and the tool follows.

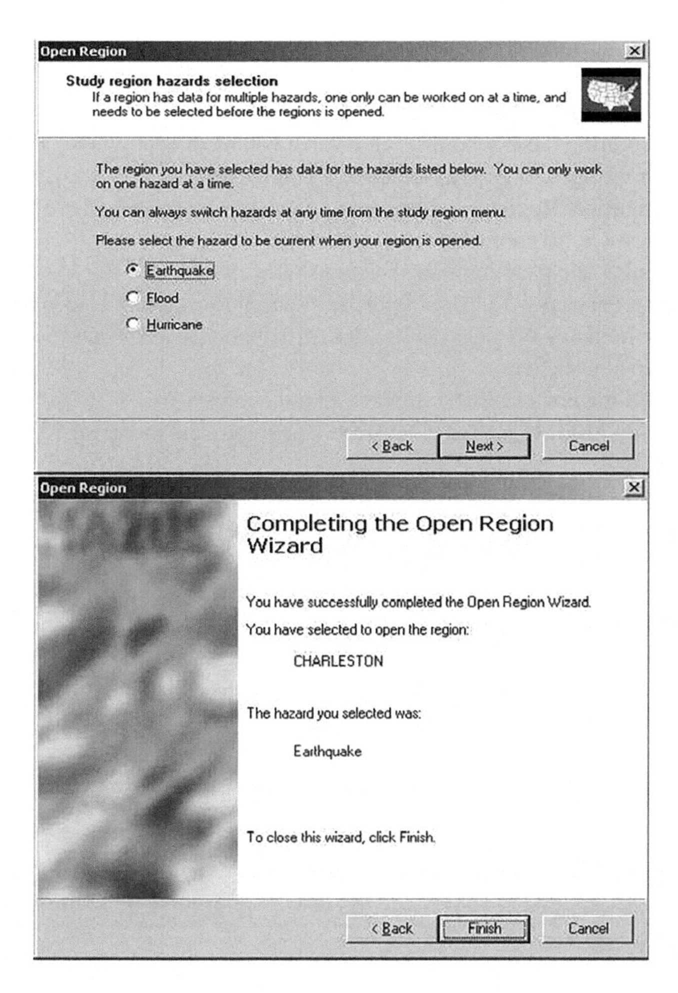

FIGURE 3.4 Step 2(b): Accessing a created region (continued). (based on the Hazus-MH software).

Sahana Eden is a family of applications that provides solutions for coordination and collaboration for organizations working in disaster management. The following modules are available:

- **Requests**—Tracks requests for aid and matches them against donors who have pledged aid. The requests system is a central online repository where all relief organizations, relief workers, government agencies and campsites for displaced personnel can coordinate the supply of aid with their demand. It allows users to allocate the available resources to fulfil the demands effectively and efficiently.
- **Volunteers**—Manage volunteers by capturing their skills, availability and allocation. The volunteers system keeps track of all volunteers working in the disaster region. It captures not only the places where they are active but

also captures information on the range of services they are providing in each area. Features include:

i. Registering ad hoc volunteers willing to contribute
ii. Capturing the essential services each volunteer is providing and where
iii. Choosing skill and resources of volunteers

- **Organization Registry**—Lists “who is doing what and where”. Allows relief agencies to coordinate their activities
- **Hospitals**—Helps to monitor the status of hospitals
- **Missing Persons**—Helps to report and search for missing persons
- **Shelter Registry**—Tracks the location, distribution, capacity and breakdown of victims in shelters
- **Map**—Situation awareness and geospatial analysis
- **Ticketing Module**—Master Message Log to process incoming reports and requests
- **Person Registry**—Central point to record details on people
- **Disaster Victim Identification**—Disaster victim identification
- **Administration**—Site administration
- **Document Library**—A library of digital resources, such as photos, signed contracts and office documents
- **Delphi Decision Maker**—Supports the decision-making of large groups of crisis management experts by helping the groups create a ranked list
- **Incident Reporting**—Incident reporting system
- **Budgeting Module**—Allows a budget to be drawn up
- **Messaging**—Sends and receives alerts via email and SMS

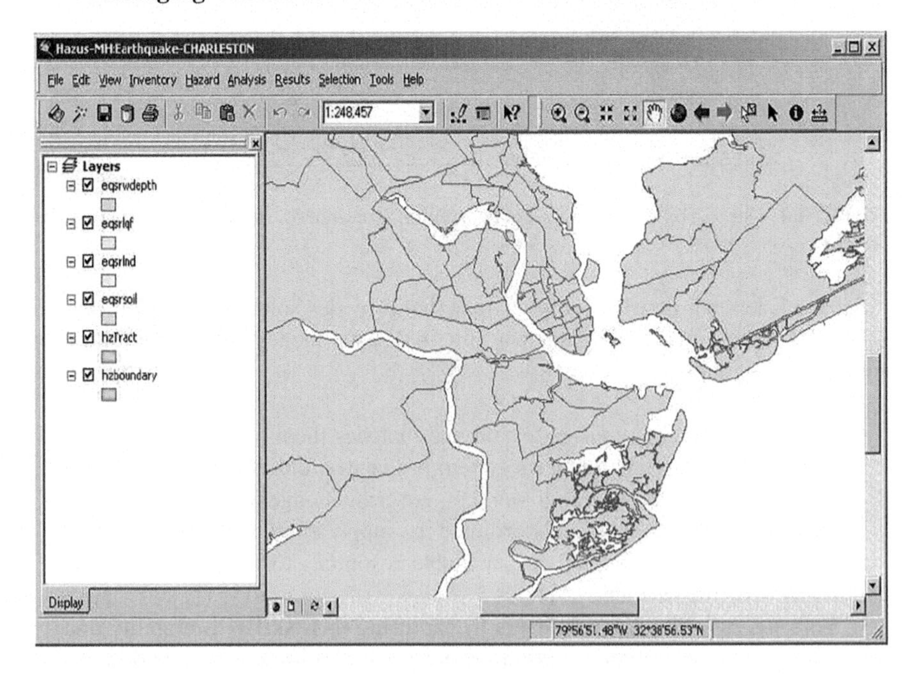

FIGURE 3.5 Step 3(a): Hazus-MH interface (based on the Hazus-MH software).

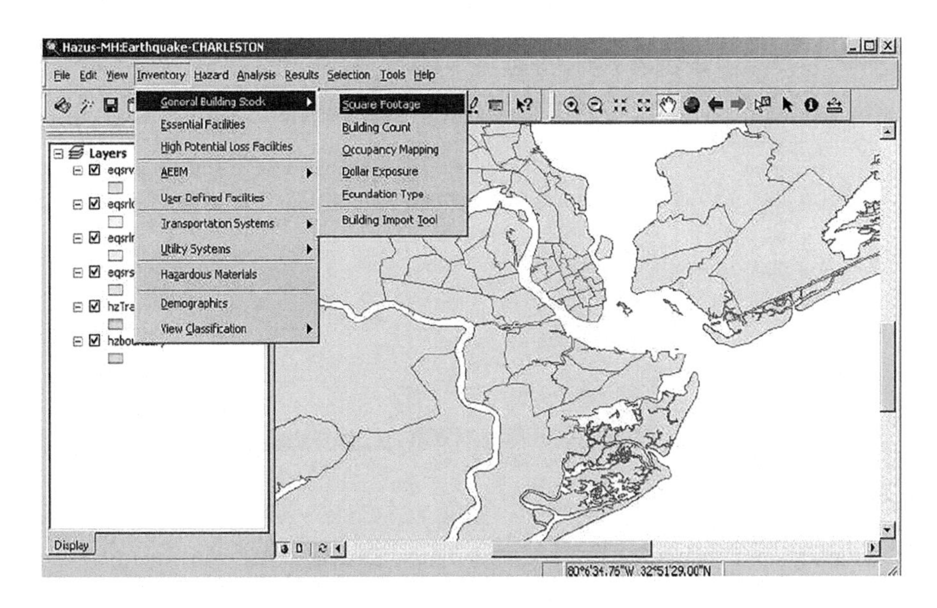

FIGURE 3.6 Step 3(b): Hazus-MH inventory menu (based on the Hazus-MH software).

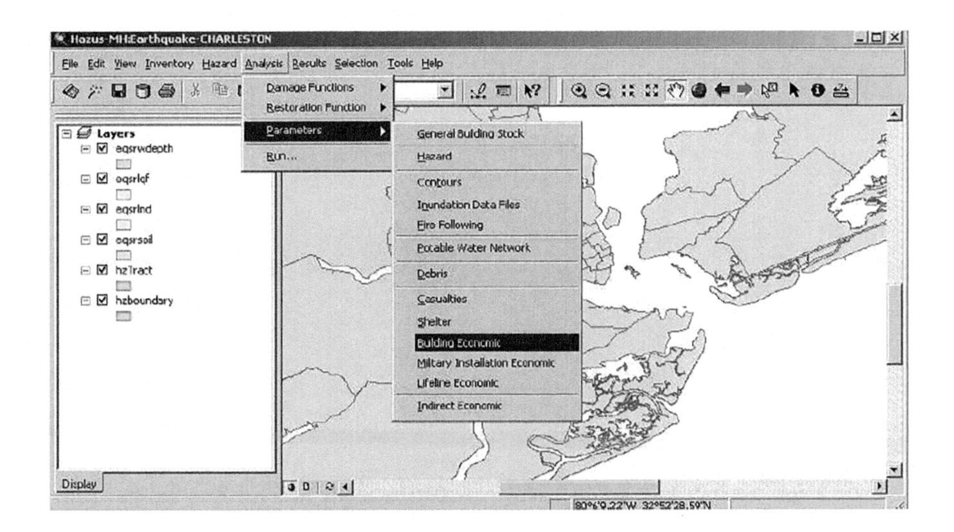

FIGURE 3.7 Step 3(c): Hazus-MH analysis menu (analytical engine) (based on the Hazus-MH software).

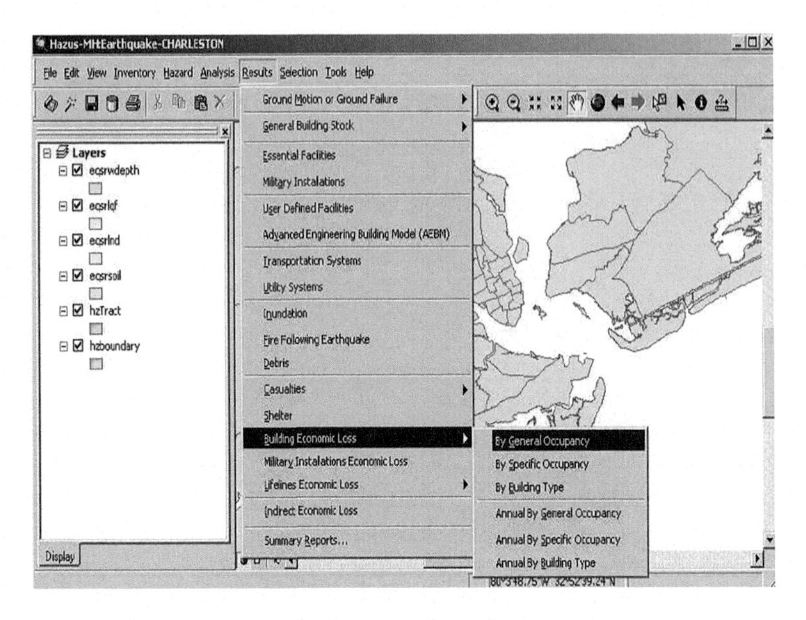

FIGURE 3.8 Step 3(d): Hazus-MH results menu (based on the Hazus-MH software).

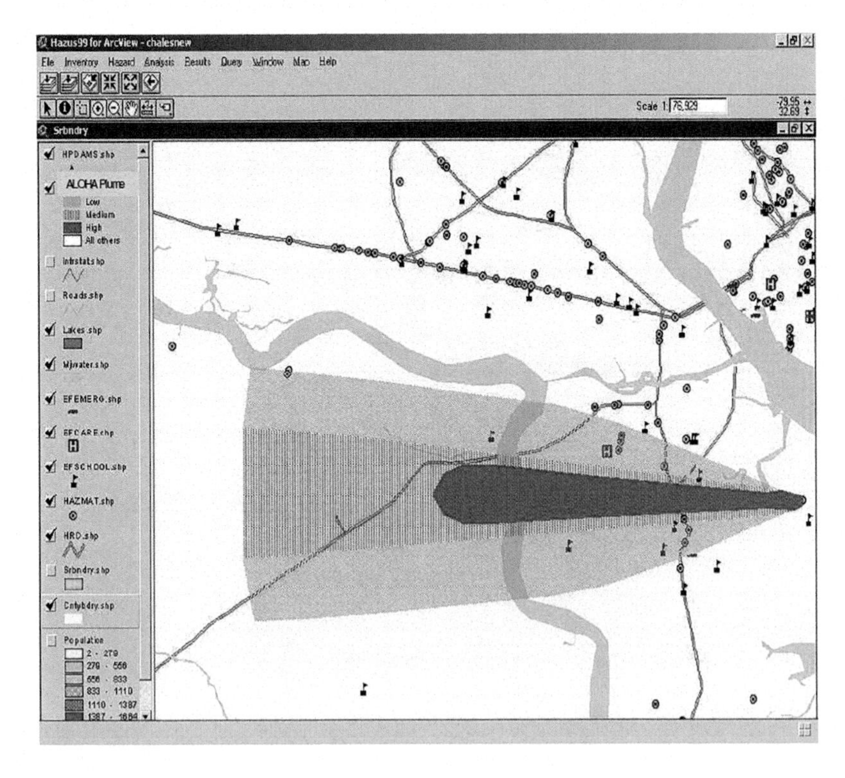

FIGURE 3.9 Step 3(e): Aloha Plume Overlaying Hazus-MH Inventory (based on the Hazus-MH software).

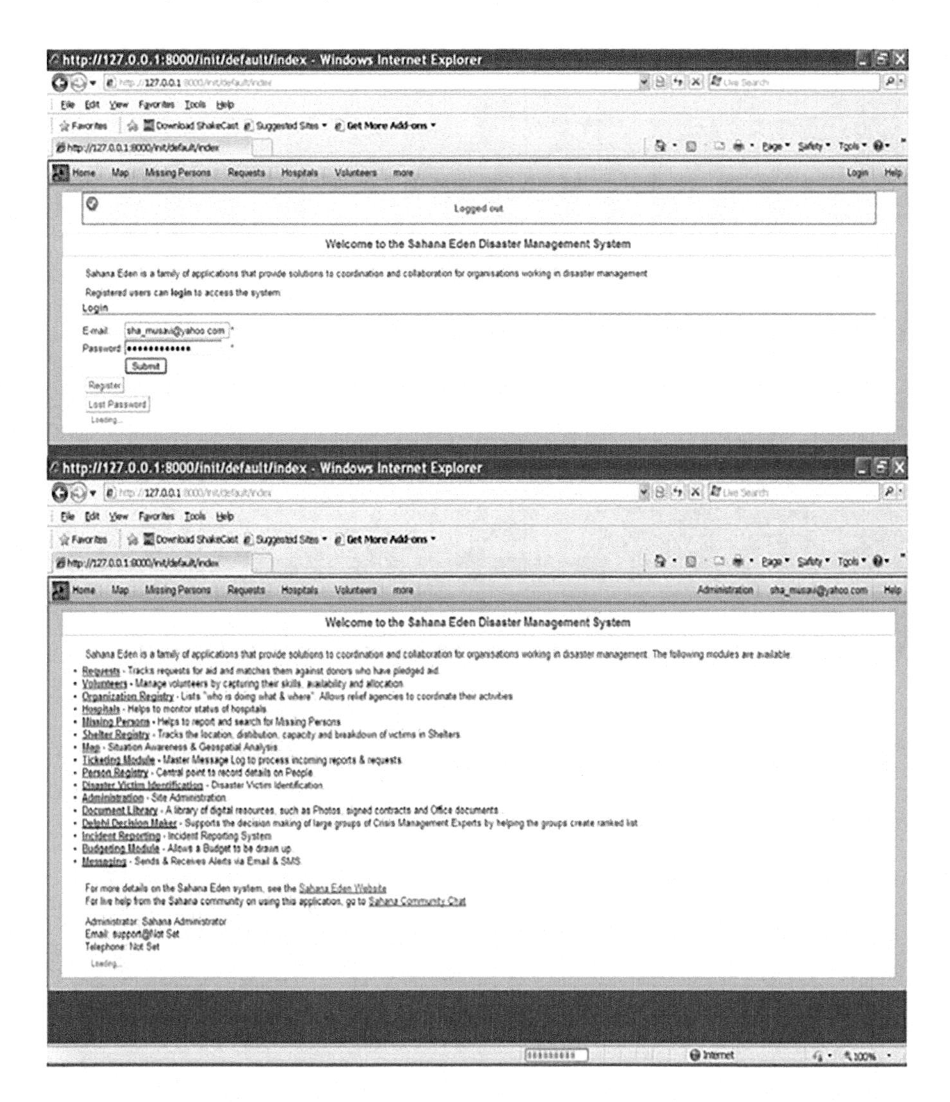

FIGURE 3.10 Sahana Disaster Management Tool used in Civionics: Login page and home page.

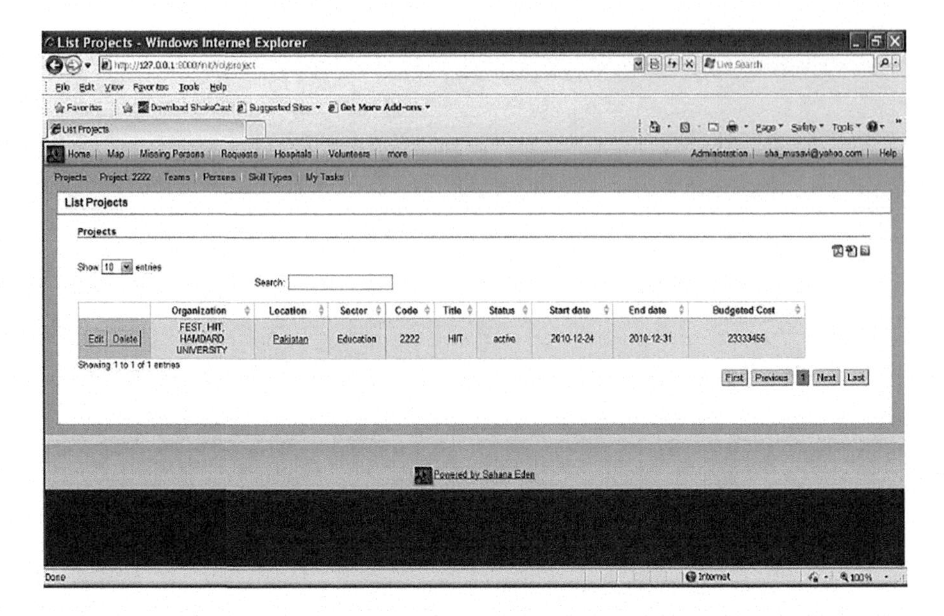

FIGURE 3.11 Sahana Disaster Management Tool used in Civionics: List Projects.

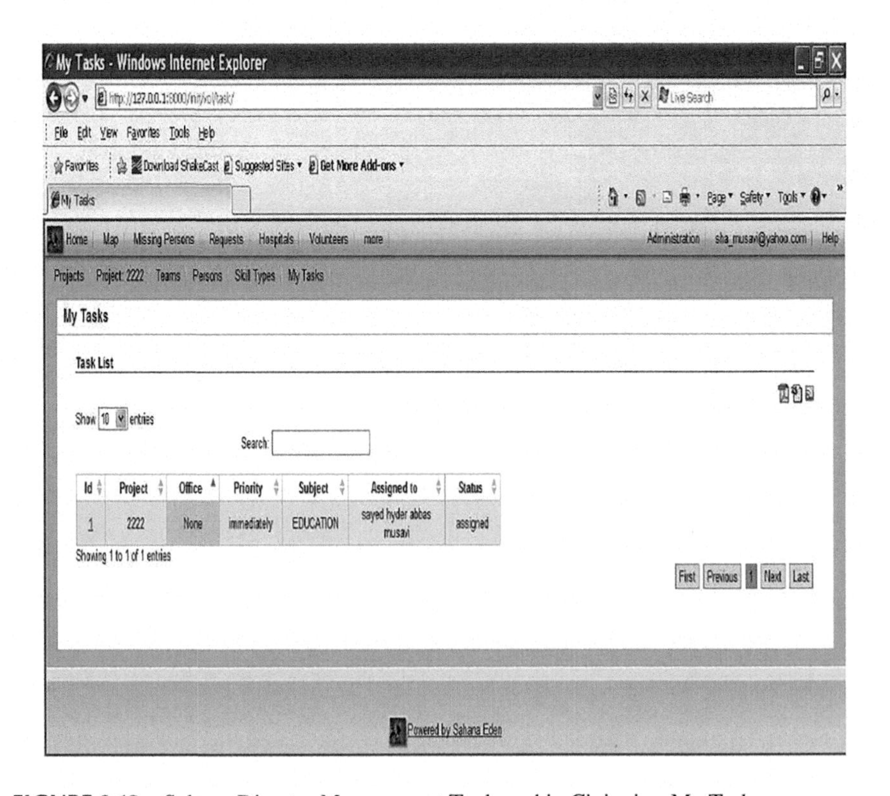

FIGURE 3.12 Sahana Disaster Management Tool used in Civionics: My Tasks.

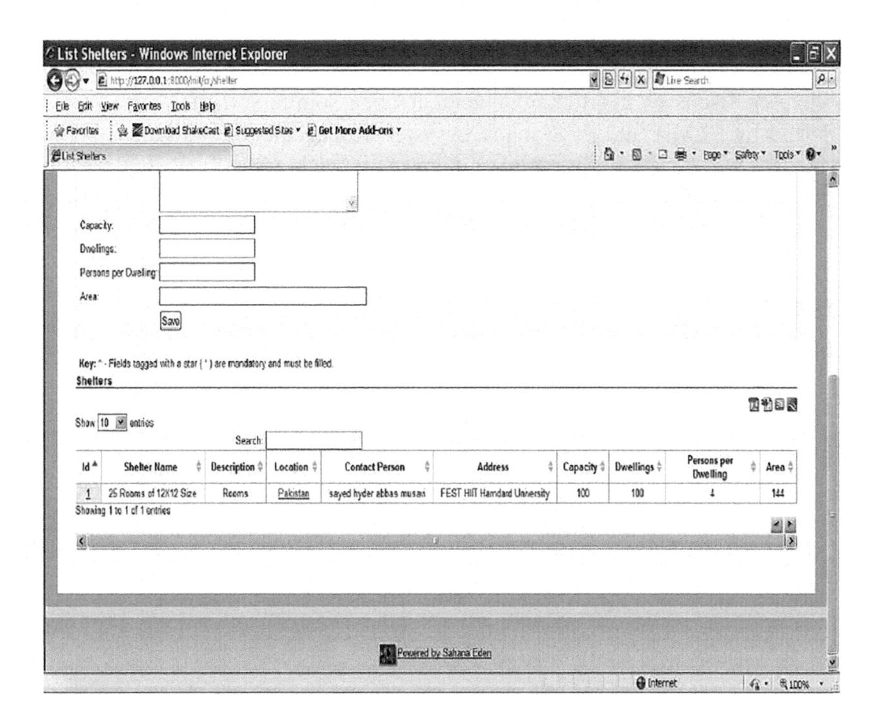

FIGURE 3.13 Sahana Disaster Management Tool used in Civionics: List Shelters.

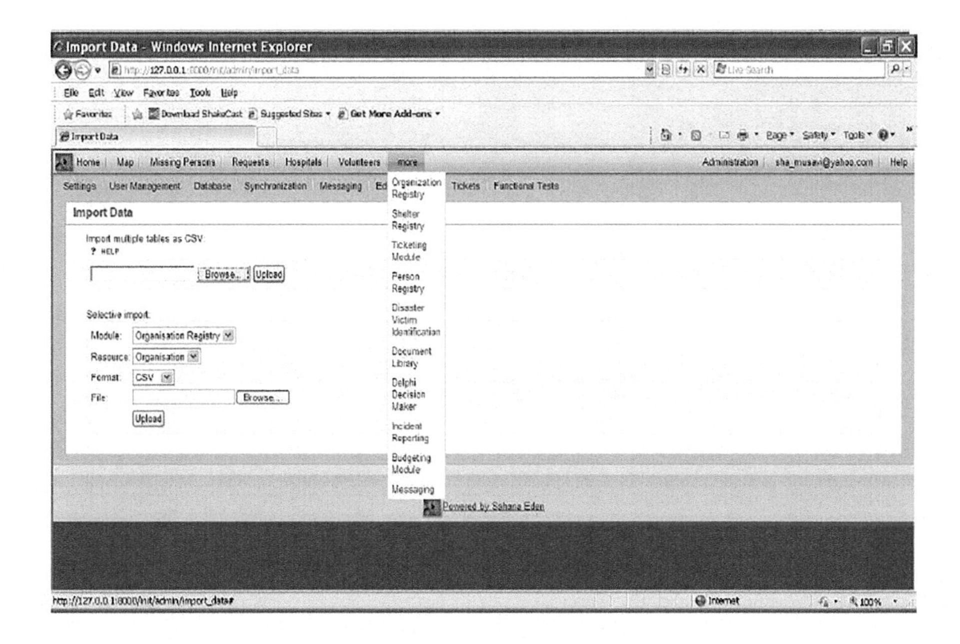

FIGURE 3.14 Sahana Disaster Management Tool used in Civionics: Import Data.

SUMMARY

This chapter discusses disaster management open source software tools, such as Hazus-MH by FEMA and the Sahana Disaster Management Tool. Hazus is being used by states and communities in support of the risk assessment that performs economic loss scenarios for certain natural hazards and rapid need assessments during responses to hurricanes.

REFERENCE

1. Federal Emergency Management Authority: Available at https://www.fema.gov/hazus

4 Early Warning System

A Use Case Scenario

INTRODUCTION

In many ways, this chapter is one of the important contributions as it discusses the context within which this work rests. The objective of the work is to design and develop a generic model of a viable multihazard early warning system. In order to develop such an efficient system, it is important that the wheel is not reinvented, or for that matter, redundant technologies are used. Also, the generic model of the proposed Civionics Earthquake Early Warning System will be provided. The name Civionics has been justifiably suggested by a Canadian professor to connote that which facilitates CIVIlians through electrONICS means. This architecture is based on a description of technology, hardware and software (infrastructure) required for the system. In addition, the research is to leverage the abilities of available seismic sensors to develop a real-time multihazard early warning system that may be installed using realistic simulation results achieved through an OPNET Modeler to generate robust, timely and automatic warning messages and alerts for communities and installations at risk.

Pakistan is one of the most disaster-prone countries in the world. A bundle of agencies claiming to be responsible for disaster management exist in Pakistan. In this section, we shall take a brief look at these organizations. But first, a brief introduction about the country's demography is presented.

4.1 DEMOGRAPHY OF PAKISTAN

The demography of Pakistan is between longitudes 61° and 76° E and latitudes 24° and 37° N with a land area of 796,095 km^2. The country is surrounded by high mountain ranges in the north, such as the Himalayas, the Karakoram and the Hindu Kush, and a snow-covered crest, the moorlands in the west, the Salt range and Potohar Plateau, the Indus plains and the Balochistan Plateau. Iran is at its western border, Afghanistan is in the north-west, China in the north, India in the southeast and the Arabian Sea to its south. The Pakistan Meteorological Department has installed seismic observation stations in some cities. The country map as depicted in Figure 4.1 shows such a network [1].

Both natural and man-made disasters are frequently hitting Pakistan. Floods, earthquakes, landslides, cyclones, droughts, terror attacks and blasts are examples of disasters that the country is facing. Almost 85,000 people were killed and more than 4 million people were affected as a result of the October 8, 2005 earthquake in north Pakistan.

FIGURE 4.1 Weather Observatory Stations installed by the country's Met Office. (Pakistan Meteorological Department [PMD], April 11, 2010.)

4.2 MAJOR DISASTERS AND THEIR IMPACT IN PAKISTAN

Table 4.1 takes a brief look at the major disasters the country has faced. There appear to be numerous civil and military agencies working in the country to address disaster issues. However, the country lacks an effective unified multihazard disaster warning and response system to help reduce the loss of life and livelihood resulting from natural and man-made hazards. It is therefore the need of the hour that a multihazard early warning system should be developed on scientific parameters.

4.3 ORGANIZATIONS WORKING FOR DISASTER MANAGEMENT IN PAKISTAN

The lack of political stability leads naturally to an unfocused and almost non-committal approach to disaster management in Pakistan. The natural consequence is a whole host of agencies all set up in origin to perform the same role, but who have, over time, lost their focus. Whilst it is not within the scope of this work to discuss the reasons for this loss of focus, it is pertinent to name these here.

Following is a list of organizations that are responsible for disaster management in Pakistan:

- Emergency Relief Cell (ERC)
- Pakistan Meteorological Department (PMD)

TABLE 4.1
History of Natural Disasters in Pakistan

Disaster Type	Date	Affected	Death Toll
Earthquake (Quetta)	May 31, 1935	150,000	60,000
Earthquake/Tsunami (Makran)	November 27, 1945	13,000	4000
Flood	1950	75,000	2900
Windstorm	December 15, 1965	130,000	10,000
Flood	August 1973	4,800,000	1500
Earthquake (Northern Areas)	December 28, 1974	97,000	5300
Flood	August 2, 1976	5,566,000	4200
Flood	June 1977	1,022,000	10,354
Flood	July 1978	2,246,000	3100
Flood	August 1988	1,000,000	7600
Extreme temperature	June 11, 1991	1200	961
Flood	August 9, 1992	6,184,418	7800
Flood	September 1992	12,324,024	1334
Windstorm	November 14, 1993	2000	609
Flood	July 22, 1995	1,255,000	5600
Flood	August 24, 1996	1,186,131	2300
Flood	March 3, 1998	75,000	1000
Drought	March 2000	2,200,000	225
Earthquake (Muzaffarabad)	October 8, 2005	2.5 million	78,000
Flood	July/August 2010	20,000,000	3500

Source: https://en.wikipedia.org/wiki/List_of_natural_disasters_in_Pakistan

- Geological Survey of Pakistan (GSP)
- Pakistan Space and Upper Atmosphere Research Commission (SUPARCO)
- National Crisis Management Cell
- The Civil Defense Department
- Federal Flood Control Cell
- National Disaster Management Authority (NDMA)

Earthquakes, floods, tsunamis, landslides, cyclones and floods are the major threats for Pakistan. The world community was struck with horror by witnessing the devastating earthquake on October 8, 2005 in Kashmir and its adjoining areas in Pakistan. If an effective early warning system was in place in the Indian Ocean region, the impact of losses sustained by lives and property would have not been so wide and extensive. The Government of Pakistan approved the establishment of a National Disaster Management System in Pakistan on February 1, 2006, which is now active in dealing with the national issues pertaining to disaster preparedness, planning, response and recovery under the umbrella of the National Disaster Management Authority (NDMA) [2]. NDMA is also striving to stretch its base to the four provinces of the country. On March 27, 2006, UNESCO held a session in Bonn, Germany with Pakistan also being one of the participating countries. The session ended with a commitment by developing countries for the implementation of plans in their respective areas to install early warning systems and disaster management capacity building measures. In view of such sessions of UNESCO, the United Nations International Strategy for Disaster Reduction (ISDR) announced a seven-point agenda for promoting National Capacities for which the consortium promised to provide funding. Chaudhry [3] submitted an estimate of US$38.25 million for the development of an early warning system. However, the document does not speak of any detail about the technology and validation of the proposal in terms of its technical success.

Early warning systems are now operational in some countries of the world. Nevertheless, the day-to-day innovations in science and technology are producing new thoughts for treating the disaster management issue to be solved through an engineering mind-set. And one of the unique elements of this study is the design of the early warning system from this engineering/ICT perspective. The role of an efficient and robust ICT system is essential for the reduction of response time and for augmenting the meaningful usage of resources. There seems to be an increasing feeling among the general public all over the world for informing them ahead of any disaster to save lives and property. The United Nations has also applied stress on communications interoperability rescheduling for global developing countries. The Government of Pakistan in a document has also agreed upon first responders' deficient logistic capabilities and ICT infrastructure in disaster situations. [3]. The intent of this research work thus is to explore the potential of next-generation wireless networks and to develop a robust E-line and proactive disaster management system which relies on the utilization of wireless sensor networks, IEEE 802.XX standards used in addressing the scenarios falling in categories of Personal Area Networks (PANs), Metropolitan Area Networks (MANs) and Wide Area Networks (WANs). It is proactive due to its ability to deal with early warning alerts to be disseminated on time.

4.4 CIVIONICS EARLY WARNING METHODOLOGY

4.4.1 The Multihazard Approach

There appears to be an increasing thrust among all governments, corporations, groups and individuals for an advanced multihazard early warning system to prepare themselves for the impending threats so as to have advance time to save lives and property. Pakistan, too, requires proactive systems because these save lives and reduce first responder costs. In view of such a need, we propose architecture for the proactive stage of the Civionics system as appeared in Figures 4.2 through 4.5. While designing an early warning system, one has to focus on its important components such as how to form warning messages, the categorization of messages in terms of its severity, the process of its release, the communication channels and their interoperability. If any of these components do not meet the technical requirements in its design and development, the early warning system so designed will not be on a par with the set international requirements and thus will not receive popular public confidence and trust. Therefore, to implement such a system, the planners should consider the concepts proposed in Figures 4.6 and 4.7. The prototype of the multihazard early

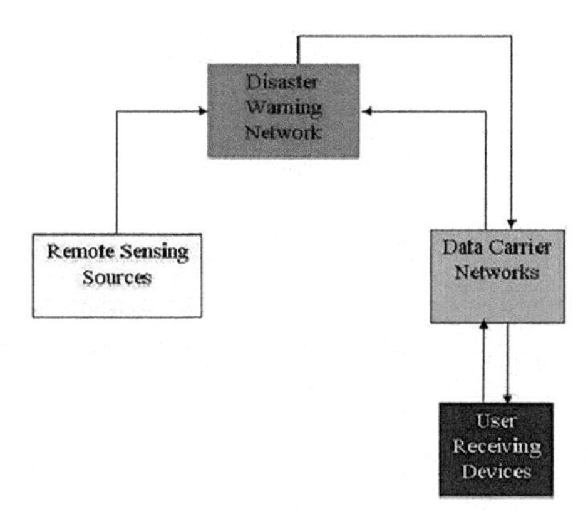

FIGURE 4.2 Civionics early warning methodology.

warning system for Civionics is shown in Figure 4.7. It is composed of identification of appropriate technology, sensors, communication links, alarm systems, computer and servers equipped with necessary software. It is important to note that except the selections for the types of sensors, the remaining infrastructure shall be somehow similar to all types of EWS used in different hazards. Furthermore, as the crux of this work is to place emphasis on the alert time between the occurrence of any disaster and the arrival of the alarm signal, our focus shall be to address in detail the disasters where this time limit is crucial, such as an earthquake. Following such a phenomenon, the planned layout of the warning stage of the system was organized as in Figures 4.2, 4.6 and 4.8. Figure 4.7 shows an image of a multihazard early warning system called Civionics. The local data processing stations are located near the installation sites of the respective sensors electronically connected with each other and also with the main warning issuing centres, which then transmit the carefully generated alarm signals to the recipients. Our aim is to communicate in advance the alarm signals within the minimum possible time as to alert on time the public, critical installations and agencies.

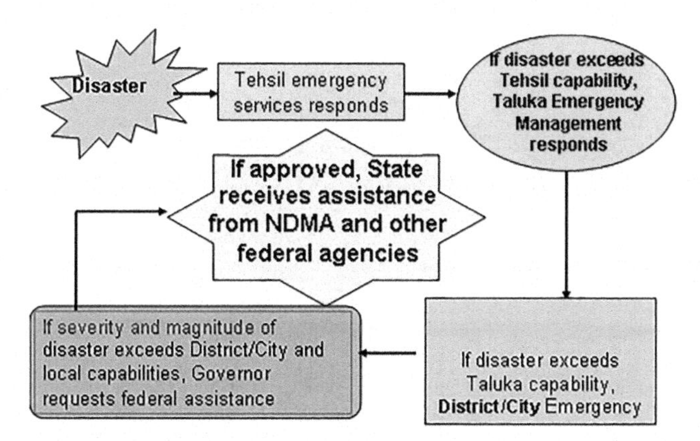

FIGURE 4.3 Disaster management hierarchies of actions in Pakistan.

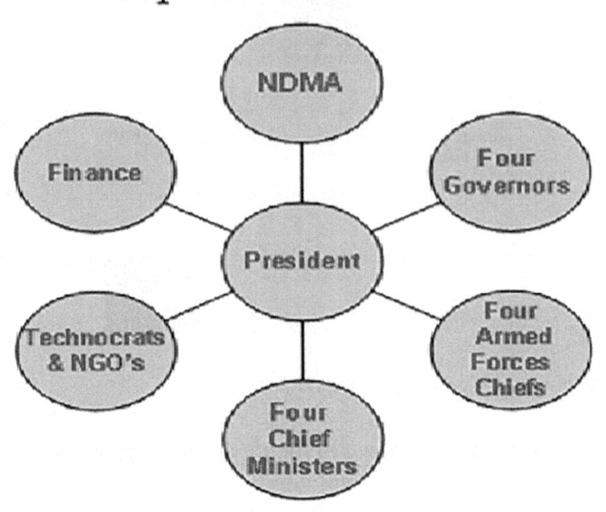

FIGURE 4.4 Proposed administration for Disaster Management Authority.

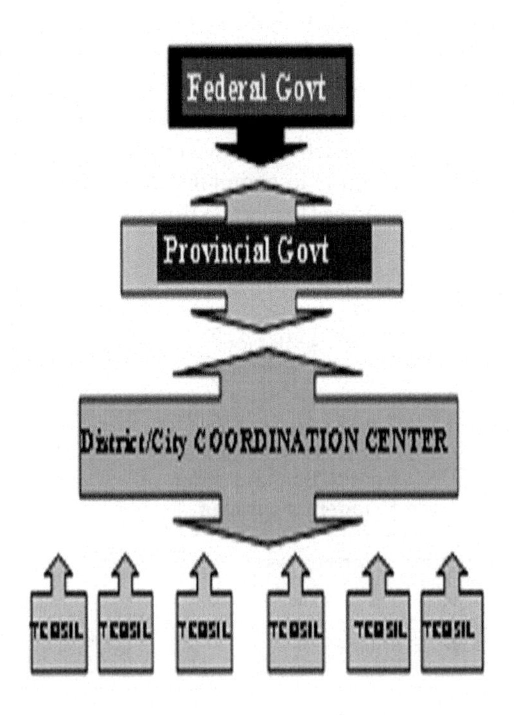

FIGURE 4.5 Disaster management top-to-bottom hierarchy.

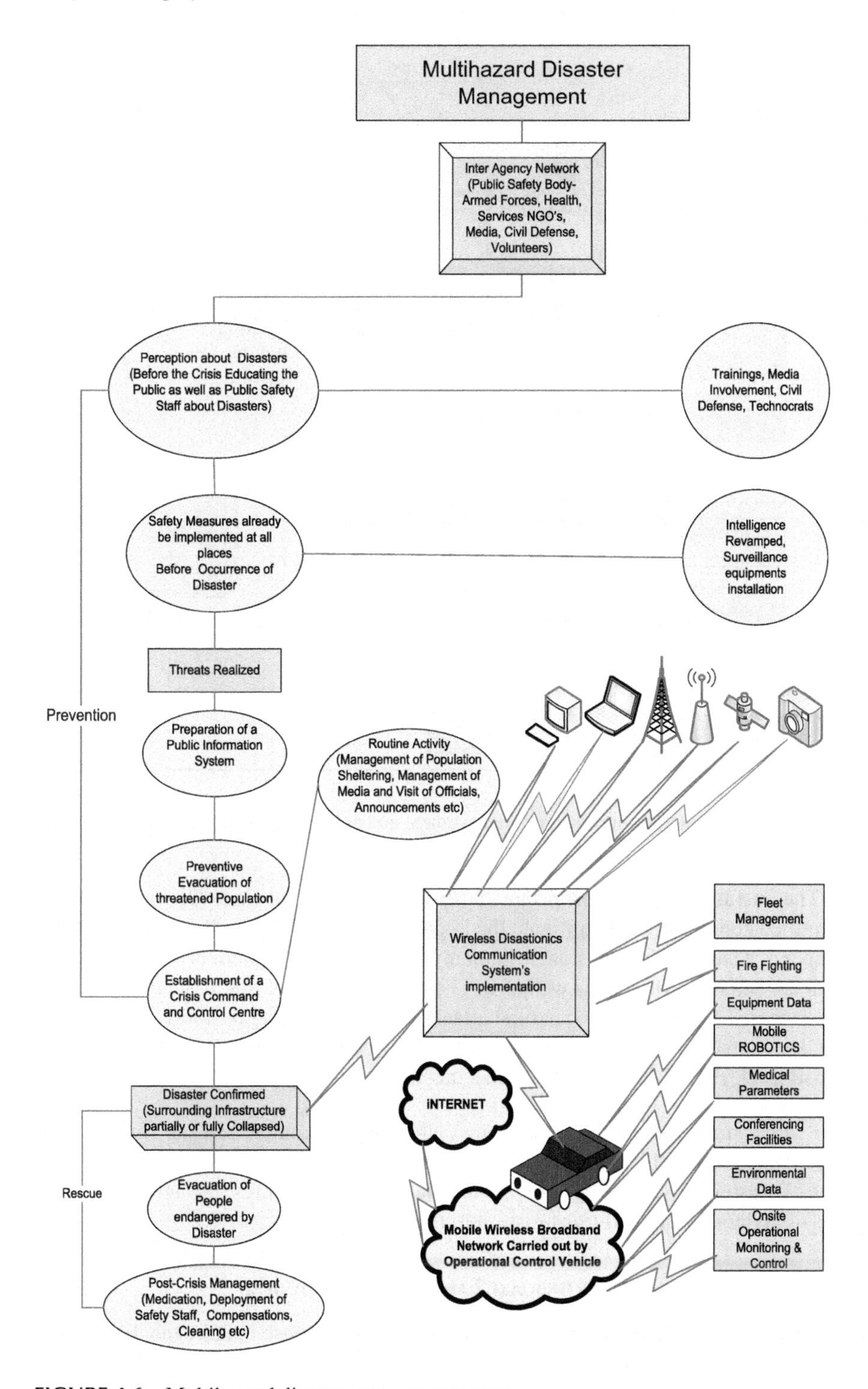

FIGURE 4.6 Multihazard disaster management steps.

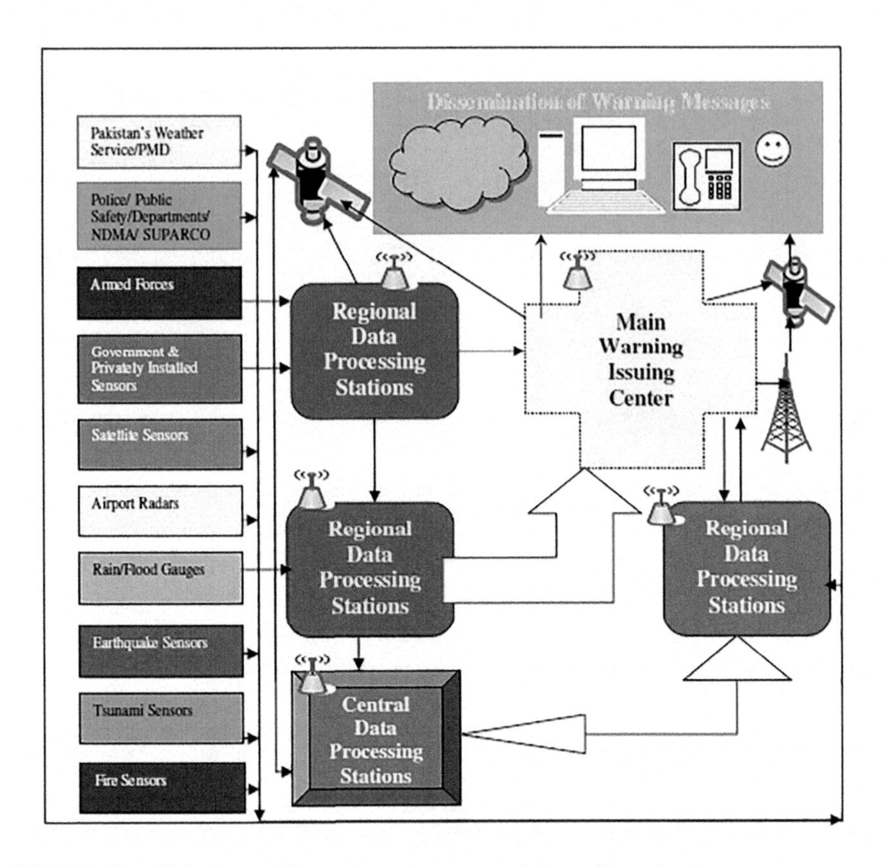

FIGURE 4.7 Civionics multihazard management and alerts dissemination concept.

The installation of appropriate hardware and software for Civionics in the identified zones shall only mitigate the impacts of disasters. In order to pre-empt the danger of a tsunami in Pakistan, a Tsunami Early Warning System is proposed in the Arabian Sea near Karachi as well as in Thatta, the interior of Sindh, while the rest of the dangers like fires, blasts, terrorist attacks, and so forth, can be addressed through the development of arrangements of ICT infrastructure for disaster at the country's Tehsil level. Figure 4.5 proposes such a hierarchy.

- Real-time data about disasters are collected from remote sensing sources
- Data collected include magnitude, location, speed and direction
- Disaster early warnings are sent to data processing centres through wireless carriers for onward transmission to users at risk

4.4.2 The Civionics Earthquake Early Warning System

The need for an earthquake early warning system is immense in Pakistan. For design purposes, two international approaches should be considered. The first approach was proposed by Nakamura in 1988 (improved later in 1996a,b) and the second by Lee

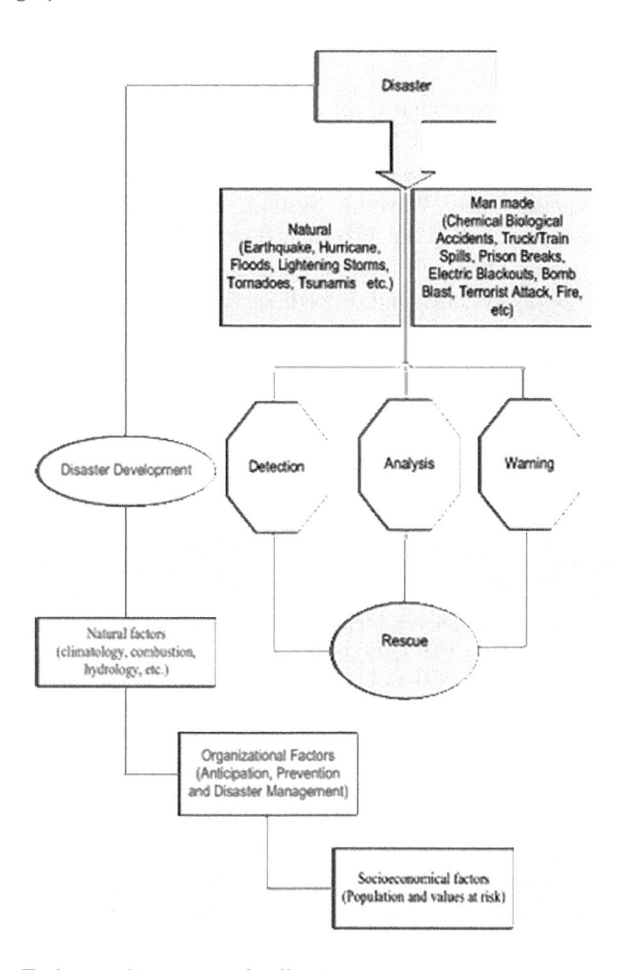

FIGURE 4.8 Early warning systems for disaster.

et al. in 1996 [4,5]. In the former approach, the seismic signals are processed locally and an earthquake warning is issued when the ground motion exceeds the triggered threshold. The latter approach is also called the array approach by virtue of which one main station processes seismic signals from an array of stations and decides whether the earthquake has occurred or not [5]. We propose the Civionics Earthquake Early Warning System (CIVEEWS) to be a regional-type alerting network deployed alongside the fault line of Pakistan which detects seismic events at the target site, and the location and size are calculated. The target side in our case is regional data processing centres responsible for the dissemination of warning alerts like the one implemented in Taiwan [6]. However, the novelty of our work is, inter alia, the inclusion of the innovative idea of Zigbee sensor nodes with WiMAX technology in these early warning networks. The CIVEEWS network not only disseminates the event information but also shuts down the operations of remote valves of industrial processes, energy generators, gas stations, railways, atomic centres, oil refineries, water supply systems, and so forth, if the need so arises.

SUMMARY

In this chapter, a use case scenario for an early warning multihazard system for developing countries is proposed. Such a system is called Civionics, which connotes that which facilitates CIVIlians through electrONICS means. Furthermore, the Civilians Earthquake Early Warning System (CIVIEEWS) network not only disseminates event information but also shuts down the operations of remote valves of industrial processes, energy generators, gas stations, railways, atomic centres, oil refineries, water supply systems, and so forth, on demand.

REFERENCES

1. Pakistan Meteorological Department. Available at www.pmd.org.pk
2. National Disaster Management Authority Pakistan. Available at www.ndma.org.pk
3. Chaudhry, Q. "Strengthening National Capacities for Multi Hazard Early Warning and Response System Phase I," Cabinet Division Government of Pakistan, Islamabad, 2006. Available at www.pakmet.com.pk/Establishment%20of%20Early%20Warning%20 System.pdf
4. Lee, W.H.K. "*Earthquake Early-Warning Systems: Current Status and Perspectives.*" United States Geological Survey, Menlo Park, CA 94025, USA. Available at http://www.cires.org.mx/docs_info/CIRES_007.pdf
5. Aldo Zollo, D.C.S., "Real Time Location for a Seismic Alert Management System-Development, HW/SW Integration, Definition and Study of Velocity Models," PhD Thesis, Universit`a di Bologna: Campania Region, Southern Italy, 2008.
6. Wen, K.-L. et al., "Earthquake Early Warning Technology Progress in Taiwan," *Journal of Disaster Research*, Vol. 4, No. 4, 2009.

5 Early Warning System Architecture

INTRODUCTION

The major earthquakes that struck the country include the Kashmir earthquake in October 2005 (M 7.6), the Quetta earthquake in 1935 (M 7.4), the Match earthquake in 1931 (M 7.3), the Pattan earthquake in 1974 (M 6.0) and the Makran coast earthquake in 1945 (M 8.0), claiming a death toll of thousands of lives and property loss of millions of dollars. This shows the country's need to invest in developing a robust earthquake early warning system.

5.1 CIVEEWS DATA COLLECTION AND IDENTIFICATION OF SEISMIC HAZARD ZONES

For achieving the goals related to designing an EWS in Pakistan, we initially suggest distributing the entire geographical area of the country into zones. In our case, the selection of zones is based on the research study report of 2007 carried out by the Pakistan Meteorological Department and NORSAR – Norwegian Seismic Array – under the title "Seismic Hazard Analysis and Zonation for Pakistan, Azad Jammu and Kashmir." This study is the result of the three-year cooperation between the Pakistan Meteorological Department and NORSAR, Norway. It was conducted by the PMD personnel Mr. Zahid Rafi and Mr. Ameer Hyder Leghari and the NORSAR personnel Dr. Conrad Lindholm, Dr. Hilmar Bungum and Dr. Dominik Lang [1]. In this report, the country was divided into 19 zones including some portions of the neighboring countries such as Afghanistan, Tajikistan, Iran and India. The division of the region into these source zones is based on the seismicity, the fault systems and the stress direction analysis. The 19 seismic zones are all having geometric shapes (polygons) and the coordinates of their corners are described in Figure 5.1.

In our research, the databases of Harvard, the International Seismological Centre England (ISC) [2], US Geological Survey (PDE) [3] and Pakistan Meteorological Department (PMD) [19], in addition to the instrumental database of USGS PDE [4], called NEIC-PDE from the Internet and the Incorporated Research Institutions of Seismology (IRIS) [5] have been used for identifying the potential seismic hazard zones in the Civionics seismic network design. Most of these databases contain hundreds of pieces of up-to-date information on earthquakes for the last 500 years. After considering the databases on the matrices of seismicity, tectonics, geology

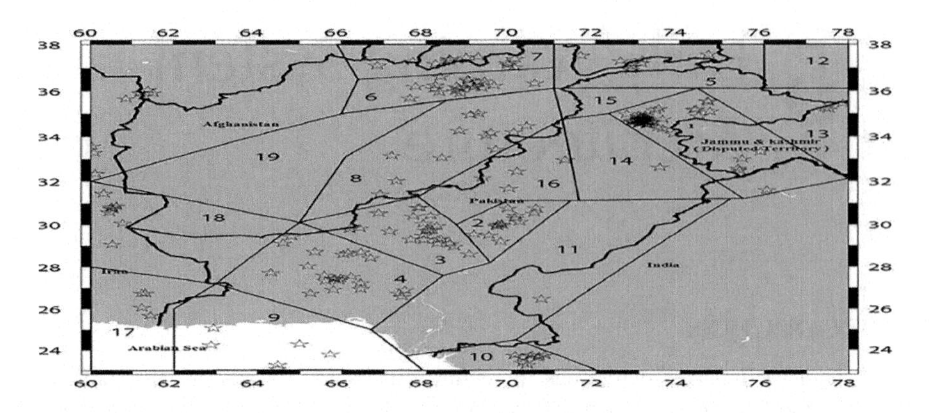

FIGURE 5.1 Seismic zones in Pakistan. (Pakistan Meteorological Department and NORSAR Norway: Islamabad.)

and attenuation characteristics of the seismic waves, the seismic hazard analysis is used to provide estimates of the site-specific design and ground motion at the site of a structure [24]. These catalogues also provide the Expected Peak Ground Acceleration (PGA) in m/s/s against an annually exceeding rate of 0.02, 0.01 and 0.002 with return periods of 50, 100 and 500 years for solid rocks in and around the different cities of Pakistan. After observations of the several wave patterns and other related information, a seismic network called the Civionics Earthquake Early Warning System (CIVEEWS) spread over the entire geographical area of Pakistan was designed. The network is composed of an MEMS accelerometer, velocity and displacement sensors connected through DSL, wireless and point-to-point radio wave communication systems. MEMS accelerometers are the most widely recommended sensors to be used for EEW systems. The network of accelerometers should be based on a distributed system with each of these sensors installed no more than 40–60 km apart along the fault lines of the country. The accelerometers will be sending the data in real time to the data processing centre. An event identification algorithm will process the individual signals received from each station to ascertain the magnitude of the earthquake based on the growth rate of the waves. If the conditions for a big earthquake are met by more than one station, the CIVEEWS network will trigger the alarm, which will be transmitted by communication technology to the conventional receivers set at public and private places and to the first responders.

According to Lindholm et al. [1], the high-seismic intensity zones in the country are the Hindu Kush region, northern areas of Pakistan and Kashmir, the north-western part of Baluchistan, the coastal areas of Pakistan (near the Makran region) and the south-eastern corner of Pakistan (RANN of KUTCH). In this report, the country was divided into 19 zones as shown in Figure 5.2, whereas Figure 5.3 indicates the seismic hazard map of Pakistan prepared for PGA for a 500-year [1] return period. However, after careful consideration of the tectonic positions in

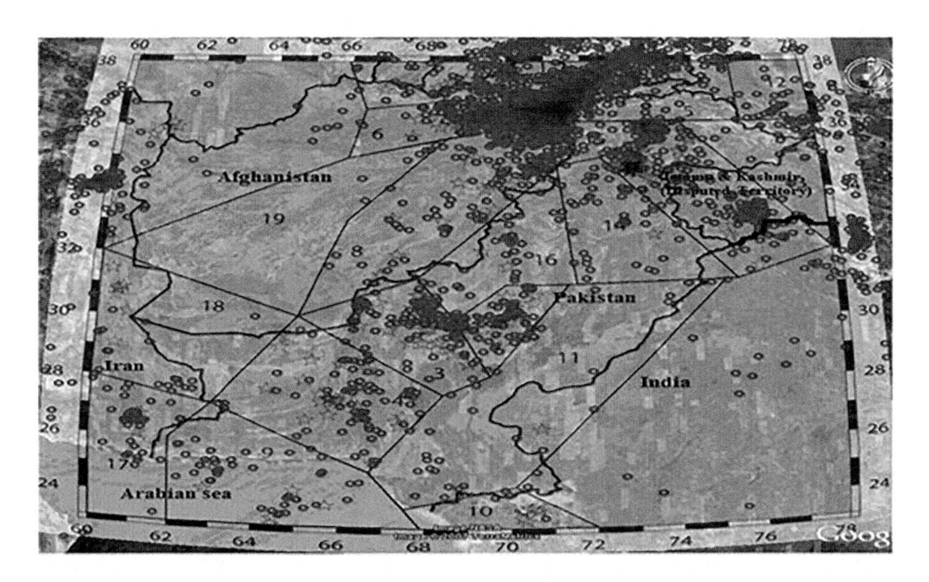

FIGURE 5.2 Pakistan's seismic map with the 19 zones overlaid in Google Earth. (Adapted from Lindholm, C. et al., "*Seismic Hazard Analysis and Zonation for Pakistan, Azad Jammu and Kashmir*," Pakistan Meteorological Department and NORSAR Norway, Islamabad, 2007.)

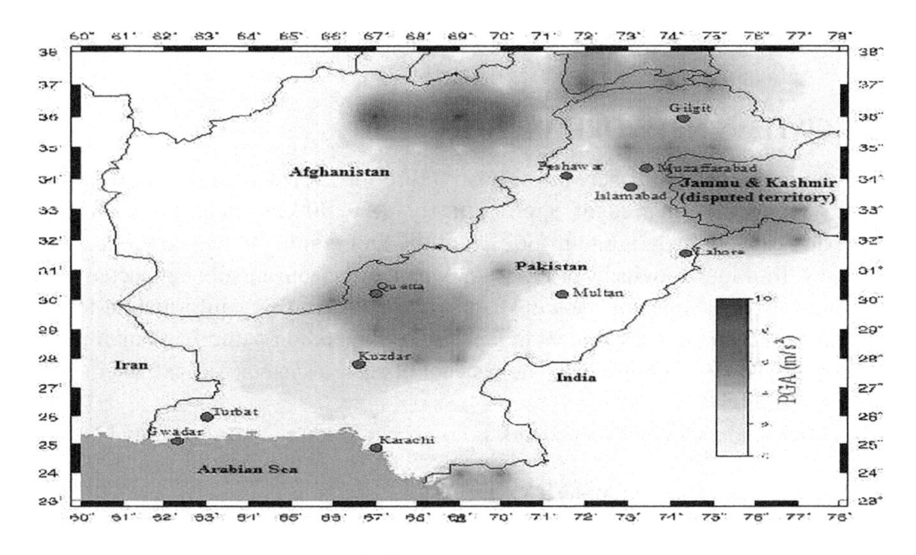

FIGURE 5.3 Seismic hazard map of Pakistan prepared for PGA for a 500-year return period. (Adapted from Lindholm, C. et al., "*Seismic Hazard Analysis and Zonation for Pakistan, Azad Jammu and Kashmir*," Pakistan Meteorological Department and NORSAR Norway, Islamabad, 2007.)

TABLE 5.1
CIVEEWS Network Designed in the High-Seismic Areas of the Country

Location	Regions	Location	Regions
35.0° N 72.3°E	Mengora	30.0°N 68.7°E	Loralai Sibi
28.2°N 69.5°E	Kashmore-Kandhkot	31.1°N 71.4°E	Lieua
27.6°N 68.4°E	Naudero	31.3°N 67.8°E	Chaman Border
27.2°N 63.0°E	Kharan Turbat	36.1°N 71.0°E	Chitral
36.1°N 76.0°E	Kashmir1	38.0°N 76.0°E	Kashmir2
25.0°N 66.7°E	Inside Sea Near Karachi	23.0°N 67.9°E	Run of Kutch
23.7°N 67.5°E	Karachi1	24.5°N 70.1°E	Umer-Kot Badin
27.6°N 68.4°E	Larkana-Sukkur	34.7°N 71.2°E	Chitral2-Afghan Border
31.1°N 71.6°E	Tobateksingh/Lieu	36.1°N 74.4°E	Baam/Kashmir
31.1°N 69.9°E	Zhob-Taunsa	28.0°N 60.0°E	Baam-Afghan Border
30.1°N 65.0°E	Afghan Chaman Border	32.0°N 60.0°E	Naseerabad Border

terms of latitudes and longitudes given in Lindholm et al. [1], it has been observed that there appears to be an overlapping of many positions. Besides, there are also geographical limitations with respect to the implementation of CIVEEWS, as some of the locations are inside the territorial jurisdictions of India, Afghanistan, Iran and India-held Kashmir. Therefore, after removing the overlaps as laid down in Lindholm et al. [1] and considering the territorial constraints, the CIVEEWS network has been designed in the following high-seismic areas of the country as shown in Table 5.1.

5.2 CIVEEWS NETWORK ARCHITECTURE

The proposed CIVEEWS network infrastructure for EEWS consists of 20 on-site stations (covering an area of approximately 90 × 90 km each) for each data processing centre. There are four for each of the four command and control centres situated within four provinces of Pakistan, which are electronically connected with the country's other alerting systems to provide early warning information to the population and the first responders in the case of impending natural and man-made disasters as shown in Figure 5.4. The concept of such a system was extracted from Figures 5.5 and 5.6.

In our case, the CIVEEWS network is composed of three main sections:

A. CIVEEWS Onsite Stations (COS)
B. CIVEEWS Data Processing Stations (CDPS)
C. CIVEEWS Command and Control Centres (CCCC)

Furthermore, for design considerations, the CIVEEWS infrastructure is divided into four regions depending on the locations of COS inside CDPS. Four CCCC representing the four provinces of the country are designed, each one bearing

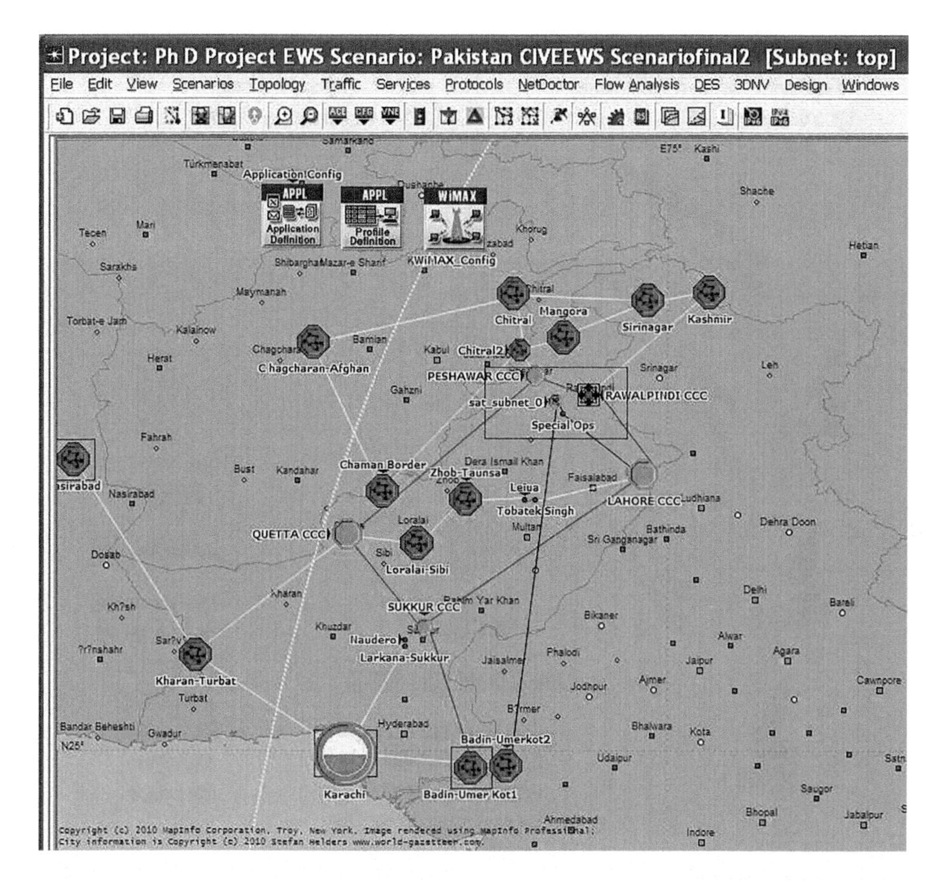

FIGURE 5.4 Network of CIVEEWS data processing stations designed in OPNET with subnets depicting four command and control centres in the country and CDPS (based on OPNET software).

the responsibility of communicating with their respective CDPS, which in turn is electronically connected with the COS. While the number of COS within a designated area varies, nevertheless an effort has been made to restrict each region within a geographical coverage of ~90 × 90 km. By approximating this range, we mean to cover a wireless communication range of any WiMAX BS, Wi-Fi devices or Zigbee nodes. Therefore, the network is distributed in multiple subnets connected with the respective CDPS for onward transmission of real-time seismic information to CCCC. Keeping in view the formats of the seismic chains in the country, the 20 COS were deployed along ellipses with a major axis parallel to every country's chain as shown in Figure 5.7.

The CIVEEWS Onsite Stations (COS) are equipped with two types of instrumentations. The northern region of the country is equipped with Colibrys SF3000L MEMS accelerometer sensors [7] at some stations and silicon design devices utilized in ESS-1221 [8] sensors at a few other stations. The former sustains

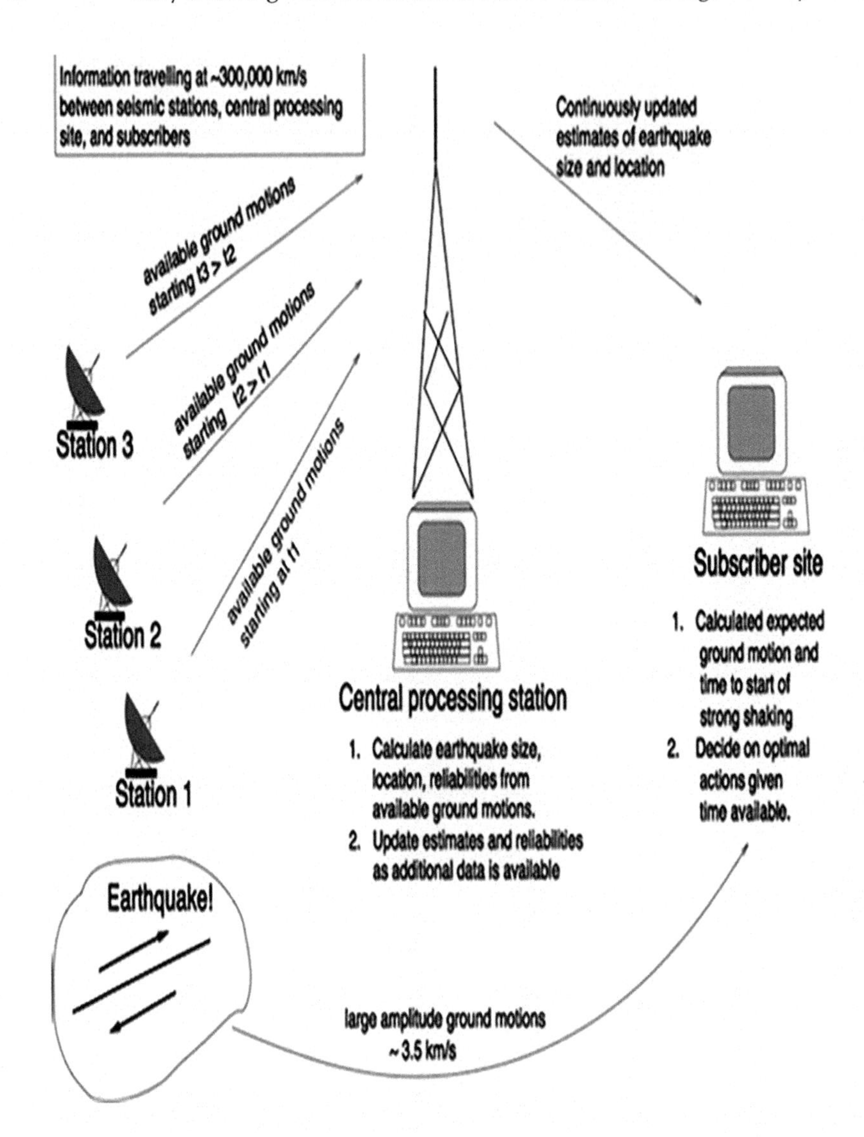

FIGURE 5.5 The concept of seismic stations deployment. (Adapted from Lee, W.H.K., "*Earthquake Early-Warning Systems: Current Status and Perspectives*," United States Geological Survey, Menlo Park, CA, 94025. Available at http://www.cires.org.mx/docs_info/CIRES_007.pdf.)

the frequency response of DC to 1000 Hz, while the latter has the capacity from DC to 400 Hz. Thus, the system has the capability of sampling up to 2000 SPS, while the southern region has been managed with Güralp Systems' instrumentation [9,10]. This distribution of instrumentation within the same country has been suggested in order to measure and compare the performance and accuracy of the two renowned systems. Equipped with solar panels possessing seven days of

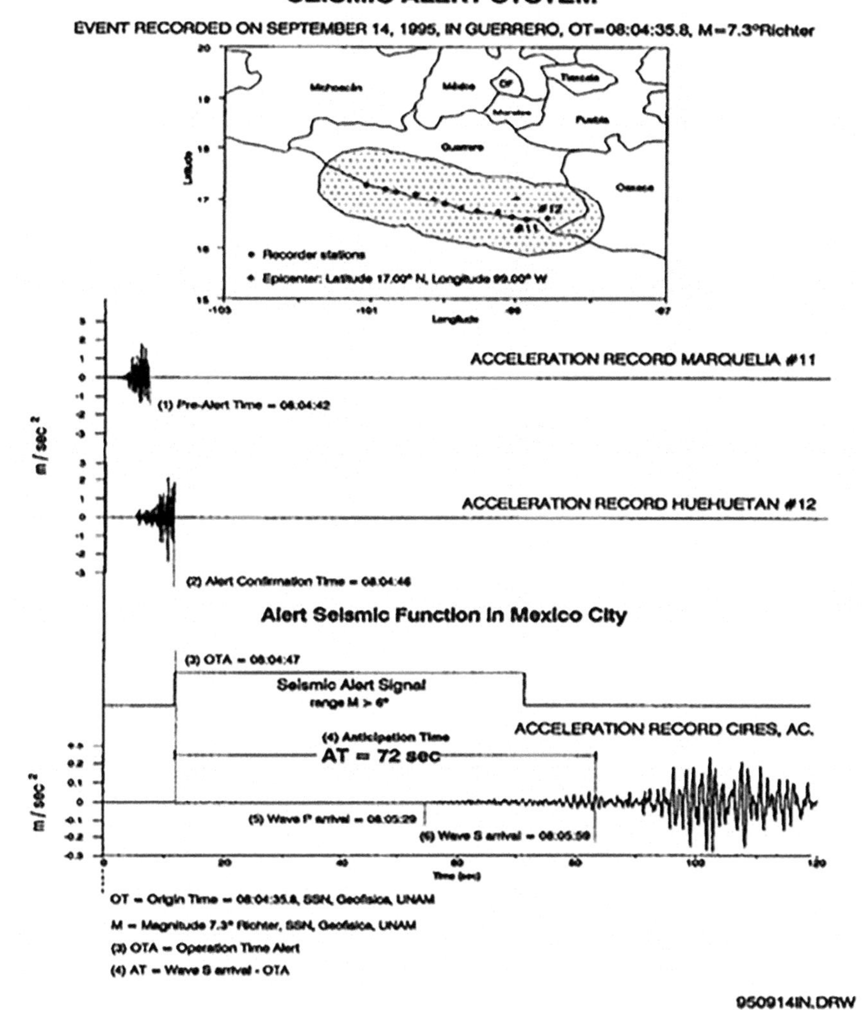

FIGURE 5.6 Time diagram of early warning advantage in Mexico City. (Adapted from Lee, W.H.K., "*Earthquake Early-Warning Systems: Current Status and Perspectives*," United States Geological Survey, Menlo Park, CA, 94025. Available at http://www.cires.org.mx/docs_info/CIRES_007.pdf.)

fully operational autonomy, each of the steel/iron boxed-COS inside the Earth is calibrated with GSM/GPRS alarm systems connected with remote environmental and industrial sensors for emergency shut off and also with data loggers. The concept of seismic on-site stations for CIVEEWS using Gularp Systems' devices is shown in Figure 5.8.

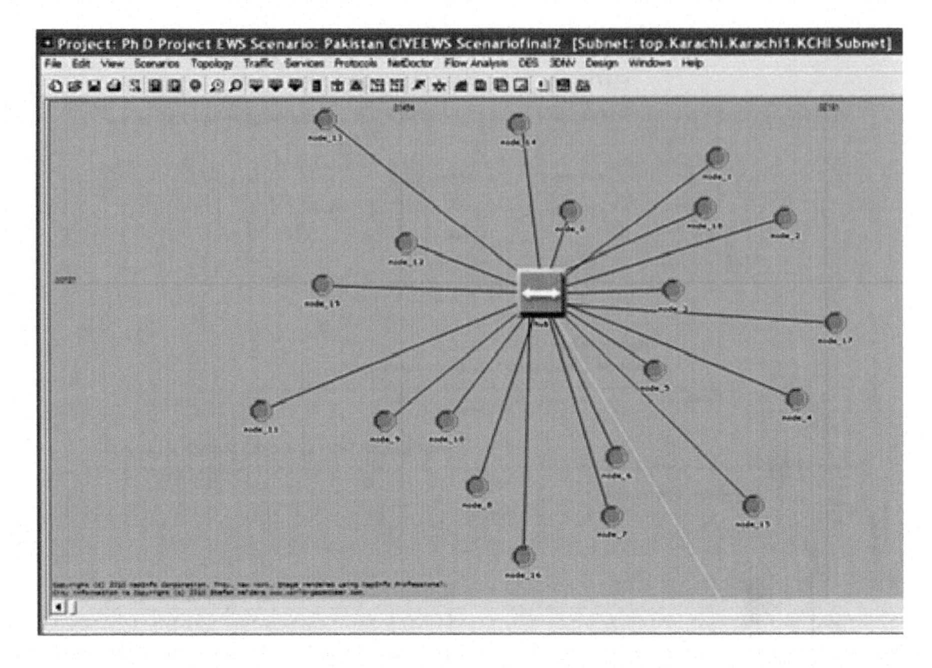

FIGURE 5.7 CIVEEWS COS deployment at one of the CDPS (OPNET software).

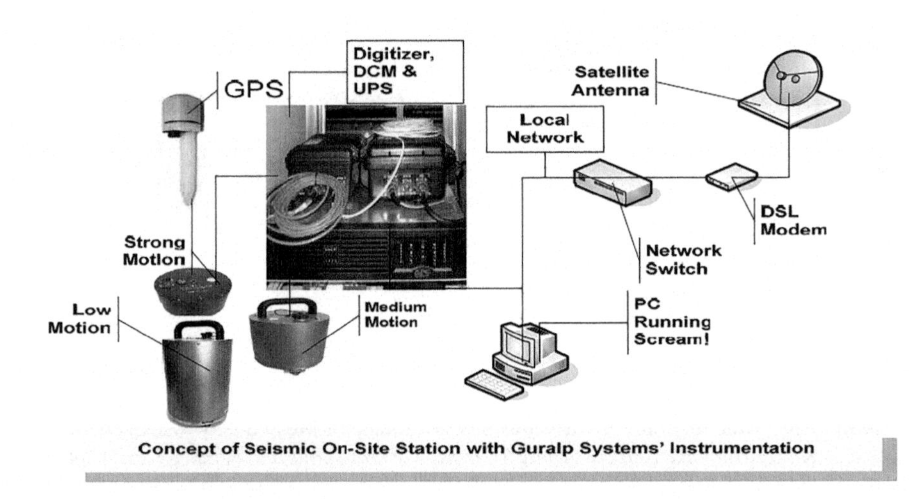

FIGURE 5.8 Setup of the seismic COS (Güralp Systems' devices).

5.3 CONCEPT OF SEISMIC WAVES

There are four types of waves involved in any earthquake – the P (primary) waves, the S (shear or secondary) waves, the L (love) waves and the R (Rayleigh) waves. The initial tremors of the earthquake are due to the P waves with short wavelengths and are less destructive than the preceding S waves. The shock of P waves is not even sensed by humans. The shaking of buildings and destruction in terms of lives and property is caused by the second wave which is expected to come after a few seconds. Every early warning system takes advantage of this time gap between P and S waves.

Hence, any EEWS is designed for initially detecting the occurrence of the P waves, which are as fast as a spark of lightning before the sounds of thunder, and the identification of the release of slower moving S waves, thereby estimating the time between the two waves as to obtain the required precision and reliability of the EWS. The speed of the seismic waves ranges from 3 to 7 km/s, whereas the information about the occurrence of the earthquake shall travel on radio technology which has the speed of light, that is, 300,000 km/s. The P waves travel at ~3.5 km/s, whereas S waves travel at about 6–7 km/s. Generally, P and S separation is 1 s for each 8 km of distance traveled [11,12]. Therefore, if an earthquake occurred at a depth of 80 km and the epicenter was at a distance of 475 km from the detector, a 60 s warning would be possible. It is apparent, of course, that if the earthquake was substantially deeper, larger in terms of magnitude and further away, an even longer warning time would follow [11,13]. Hence, the P wave precedes the S wave by the time equal to 70% of the P wave travel time to the station. The tightly spaced seismic sensors are required to be installed in the identified zones of the country. The closer to the epicenter these sensors are, the larger the time intervals will be between the arrival of the seismic and radio waves and vice versa. Clearly speaking, the data processing stations that are located close to the epicenter shall unfortunately have less time to disseminate warning messages than those that are farther from the epicenter and have larger time intervals to prepare for damage. If we deploy a dense seismic network in the earthquake source area that is capable of locating and determining the size of the event in about 10 s, we will have about 3 s to issue a warning before the P wave arrives, and about 12 s before the more destructive S waves and surface waves arrive at the city [6]. It takes very little time to send a signal from a seismic network to the city by electromagnetic waves (e.g., telephone circuit) at nearly the speed of light (300,000 km/s) [6]. Normally, an earthquake that is more than 100 km away from a city does not pose a large threat to the city, because seismic waves would be attenuated by a factor of about five in general [6]. The above strategy may work for earthquakes located about 60 km or more away from a city. For earthquakes at shorter distances (say 20–60 km), we must reduce the time for detecting the event and issuing a warning to about 5 s [6]. Because the P and S waves are important, the attributes associated with these waves are significant and therefore should be studied. There are three important attributes that are responsible for the strength of shaking. These are peak ground acceleration (PGA), peak ground velocity (PGV) and peak ground displacement (PGD).

For practical scenarios, the first integral of the strong motion signal (PGA) results in PGV, and the second integral results in PGD. These are then filtered with numerous iterative procedures using a high-pass Butterworth filter with a cut-off frequency of 0.075 Hz in order to remove low-frequency component signals attributed to noise during the first iteration. PGA, PGV and PGD are then plotted as the peak values of the three components along the *Y*-axis versus the *X*-axis, which usually is the arrival time of the P wave, as depicted in Figure 5.9 [15].

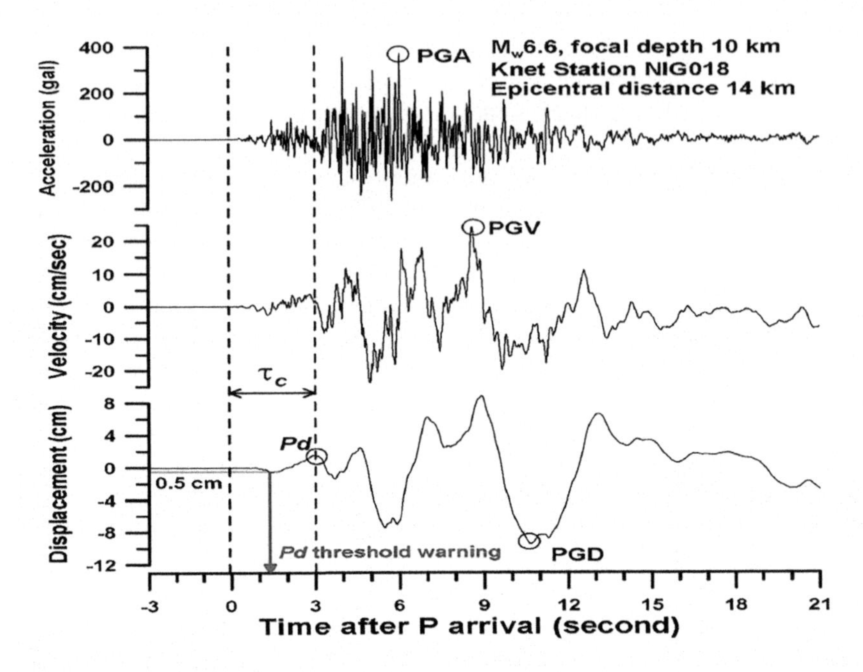

FIGURE 5.9 Relationships between displacement, velocity and acceleration. (From Wu, Y.-M., Kanamori, H., *Sensors*, Vol. 8, 2008: pp. 1–9. Available at www.mdpi.org/sensors [Last accessed on April 1, 2010].)

In 2005, Kanamori used a more modified approach than did Nakamura for earthquake early warning. He suggested computing a ratio "r" from the following Equation [13,15]:

$$r = \frac{\int_0^{\tau_0} V^2(t)\,dt}{\int_0^{\tau_0} u^2(t)\,dt} \tag{5.1}$$

Here, the numerator and the denominator are the ground motion velocity and displacement quantities, respectively, and the integration is taken over the time interval $(0, \tau_0)$ after the arrival of the P wave. Using Parseval's relation, the following relation is obtained [13].

$$\tau_c = \frac{1}{\sqrt{\langle f^2 \rangle}} = \frac{2\pi}{\sqrt{r}} \tag{5.2}$$

The parameter in Equation 5.2 can be used as the indication of the average period of the initial portion of the P wave, which corresponds to the P-wave pulse width becoming larger with the earthquake magnitude and is used to calculate the event magnitude.

Wu and Kanamori [15] further conclude that one possible approach for faster warning is to monitor peak displacement (Pd) and issue a warning as soon as it has exceeded 0.5 cm. In the 2007 earthquake in Japan, at the nearest stations (14 km), the threshold value of Pd = 0.5 cm was reached at 1.36 s from the arrival of the P wave. If we issue a warning at a threshold of Pd $\leq$ 0.5 cm, a warning will be issued at 1.36 s after the P arrival and several seconds before the occurrences of PGA and PGV. This type of early warning approach will become effective, especially for close-in sites where warnings are mostly needed [15]. The virtual seismologist method [25] and the neural network-based PreSEIS methodology [26], depend on the cumulative absolute velocity (CAV) approach, which are the other means currently under investigation in California and Switzerland to foretell seismic source and ground motion.

5.4 SELECTION OF A DATA LOGGER

The selection of a data logger for CIVEEWS is a dynamic and not an easy task. For the purpose of getting reliable results, considerations like the ability of a digitizer to measure the smallest voltage increment (LSB), the dynamic range, maximum excursion and selectable gain have been taken into account in selecting the data logger. If an A/D converter [9,10] has the ability to measure the LSB of amplitude A, then the noise power is given as $p = (A^2/2)$. If M is the number of bits of a converter, the sine wave power can be calculated as $(2^{M-1} \times A)^2/2$. The dynamic range R can be calculated as: $R = (A^2/12)/((2^{M-1} \times A)^2)/2$. After some data input, $10 \log R = 6.02 * M + 1.76$ dB.

A perfect 24-bit converter has a dynamic range of 146 dB.

The CMG-24 [9] model of Güralp Systems, as shown in Figure 5.10, which has GPS time cataloguing and UDP/IP connectivity with the Linux operating system running on an HP Porliant SeedLink server machine, is used for COS. CMG DM-24 has advanced features such as three or six low-noise 24-bit ADCs, an additional full-rate data channel for user signals and calibration, a low-power 32-bit DSP and ARM processor (<1 W recording four channels at 100 samples/s), multiple concurrent data rates up to 1000 samples/s, STA/LTA ratio, level, external and software triggering, UTC timestamped data from an attached GPS receiver, eight environmental channels with a 20-bit resolution (3 × mass position, temperature, 4 × user), calibration using step, sinusoid or pseudorandom broadband noise signals, 64 Mb flash memory

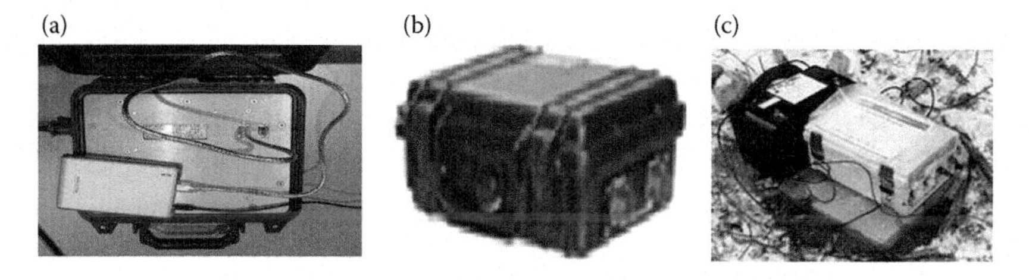

FIGURE 5.10 (a) CMD-DM24mk3 with both optional LCD display and optional FireWire interface, (b) CMG DM 24 and (c) Colibrys SF3000L MEMS accelerometer sensors (based on Gularp Systems' devices).

with fast FireWire data transfer, fully configurable using Güralp data modules and software. The DSP software on DM24 supports up to 7 cascaded filter/decimation stages. At each stage, the sample rate can be divided by a factor of 2, 4 or 5. The internal ADC output data at 2000 samples per second, so decimated data streams are available from 1000, 500 and 400 samples/s down to 1 sample/s [9,10]. On the CDPS side, Earthworm, which is the real-time seismic dispensation system, is in place to track the waveforms of 1-second packets emerging from COS.

Earthworm, on each triggered event, finds magnitude and focal mechanism estimations through real-time event detection and location tracking techniques.

DSS (data subscription service) is a packet format, widely used in strong motion projects, which enables data and statistics to be requested from a seismic installation. A DSS server is designed to handle many concurrent requests from clients with varying levels of privilege, and may prioritize requests according to their origin and urgency [10]. Güralp Systems [9] DCM data modules include a module that can communicate with installations using DSS as either a server or a client. A simple DSS server is also available, which receives requests on a network port and replies to them. Each DCM in this network runs a DSS server providing data on

- Peak acceleration levels
- RMS (root mean square) average acceleration levels
- Spectral intensity
- Magnitude of horizontal acceleration (combined N/S and E/W components)

These statistics are relayed to a central data centre once every second, where they can be used to trigger automatic warning systems. When the statistics indicate that an event has occurred, or there is a problem with the structure, DSS allows station operators to request the raw seismic data and determine the best course of action [10].

It is necessary that information about the health of the components of the whole system is guaranteed. Güralp digitizers provide a range of slow-rate auxiliary channels for reporting the system's state of health and other diagnostic information, known as multiplexed ("MUX") channels [9,10]. The number of MUX channels depends on the model and configuration of the digitizer. Generally, three channels are used to report the sensor mass position and a fourth measures the internal temperature of the digitizer. In addition to these, up to 12 MUX channels are given [9]. Some digitizers have a separate auxiliary port that can be used to access these channels [10]. Presently, the health management system of CIVEEWS can be achieved through commercially available software, but to our knowledge these are insufficient to meet the requirements. Therefore, there is a need for researchers to work on building their own SQL server-based database and component monitoring system. System reliability, accuracy and sustainability during strong seismic events are the factors to be looked into.

5.5 SIMULATION SETUP, RESULTS AND DISCUSSION

The flow chart in Figure 5.11 shows the sequence of events adopted for achieving reliable and accurate CIVEEWS results. The design of CIVEEWS requires a

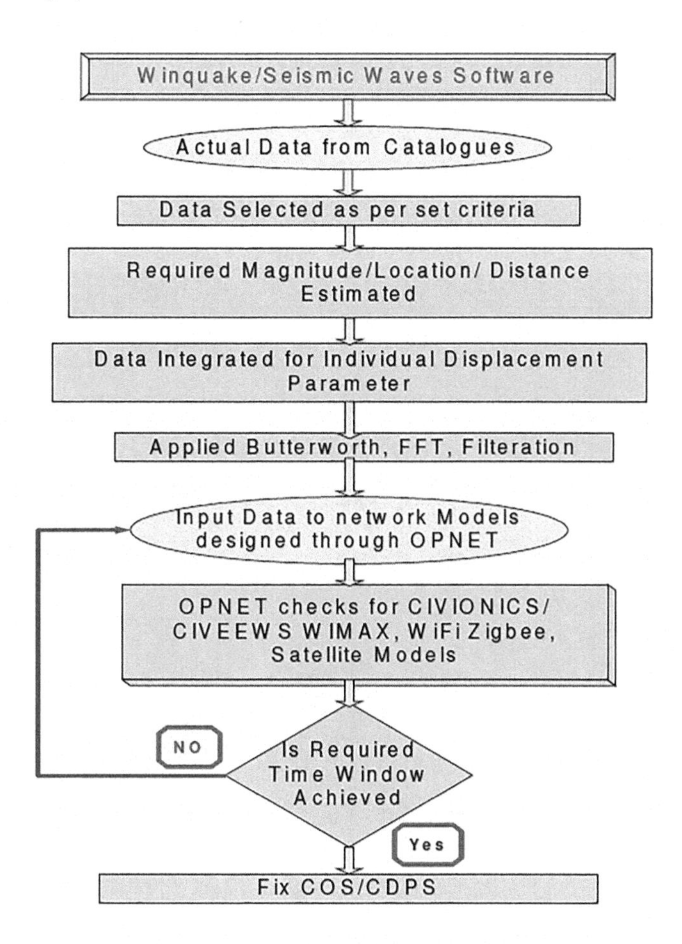

FIGURE 5.11 Flow chart of CIVEEWS simulation setup. (Based on Larry Cochrane's WinQuake Version 3.1.4b software.)

challenging task of demonstrating the prospective sites for placement of COS and CDPS. Our aim is to minimize the time window for seismic early warning messages disseminated to the population at risk. For achieving the mentioned objective, 45 data sets of the real earthquakes that occurred in Pakistan have been taken from the above-mentioned databases based on set criteria having a local magnitude greater than 5.0, focal depth <40 km and earthquakes that occurred within 50 km of the identified zones. These data sets were processed using Larry Cochrane's WinQuake Version 3.1.4b software [16] for location finding and magnitude calculation (Ms, ML, Md). Data are integrated for velocity and distance calculation, whereas the filtering operations are performed over the waves and the results are recorded. Examples of the recorded results are shown in Figure 5.12a–g.

The simulation steps shown in Figure 5.11 and the results in Figure 5.12a–g are examples from the October 8, 2005 earthquake in Kashmir, Pakistan. The selection of on-site stations for our EEWS is realized against the time gap between P and S picks at various distances of the different stations. The received data are also

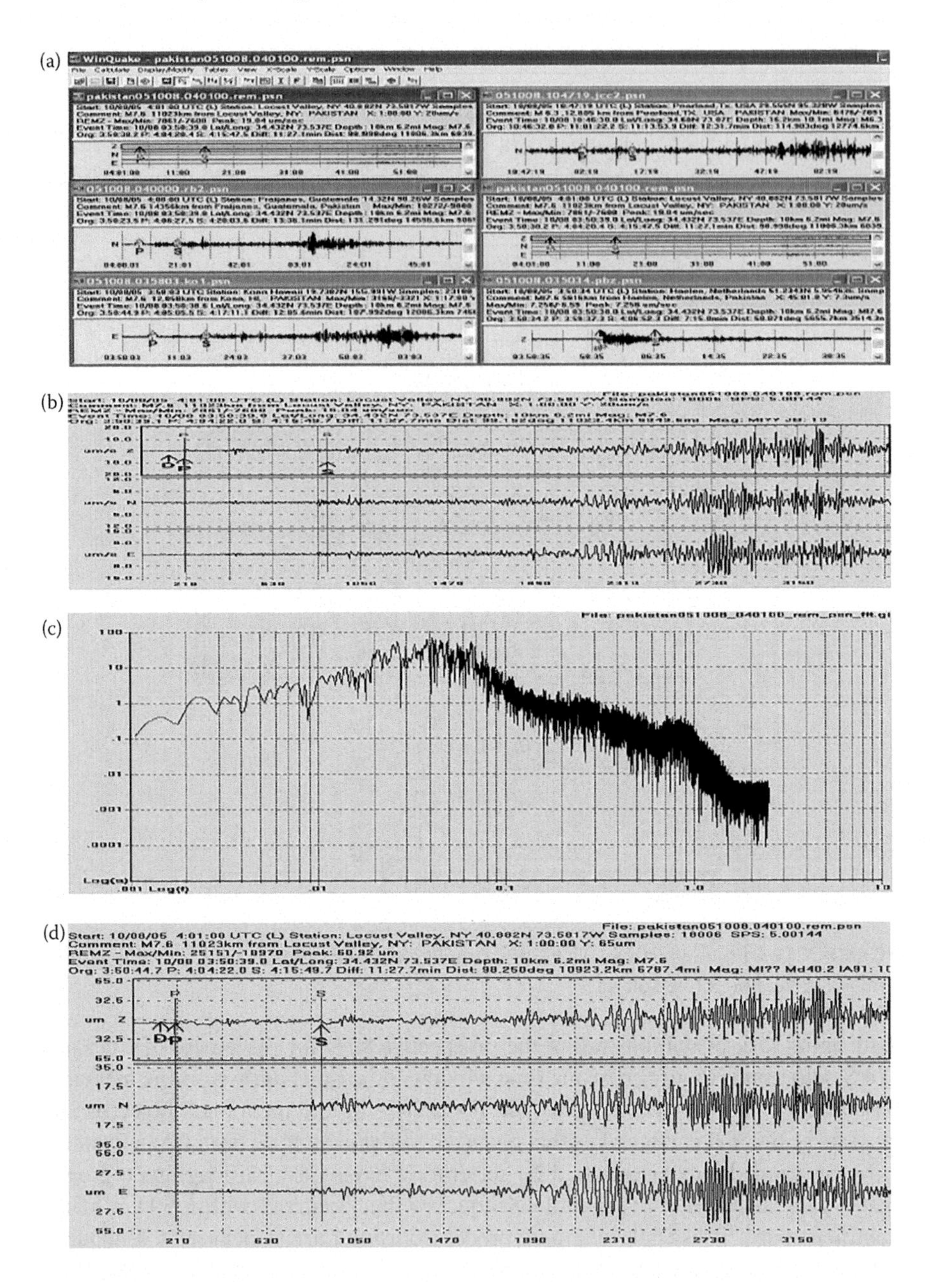

FIGURE 5.12 (a, b) Examples of Pakistan's earthquake on October 8, 2005 data collected from the IRIS catalogue. P and S indications are also given: (c) FFT of the initial wave, (d) integrated wave.

(e)

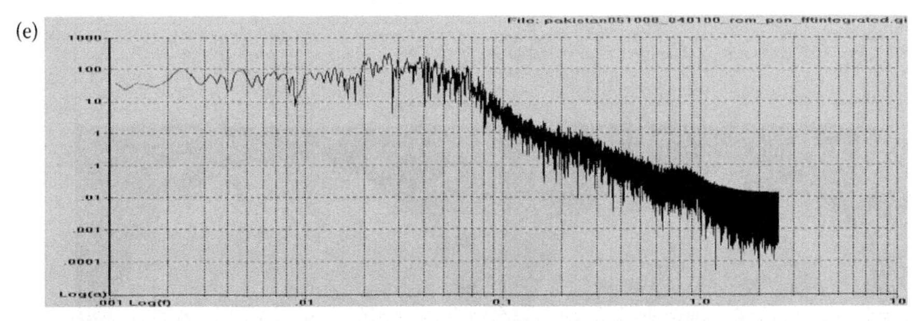

(f)

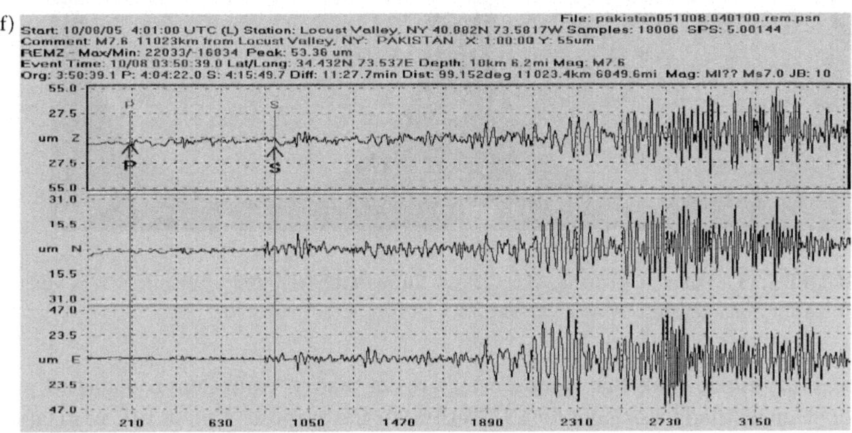

(g)

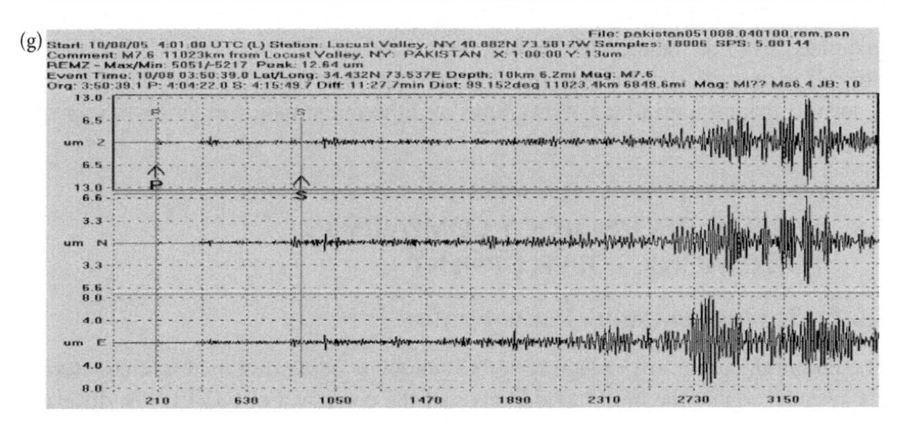

FIGURE 5.12 (Continued) (e) FFT of the integrated wave, (f) Butterworth low-pass filter at 0.075 HZ cut-off frequency with one pole and (g) Butterworth high-pass filter 0.075 HZ cut-off frequency with one pole.

analyzed for P, S Pp, Ss and other waves using the Alan Jones (Endwell, New York) software "Seismic Waves Version 2.1 level 2009.9.27" [17] as shown in Figure 5.13.

Although the tentative magnitude values and actual local magnitude values have a linear relationship and various authors have developed equations for least-squares line showing the relationship between the two, we suppose that considering and using the

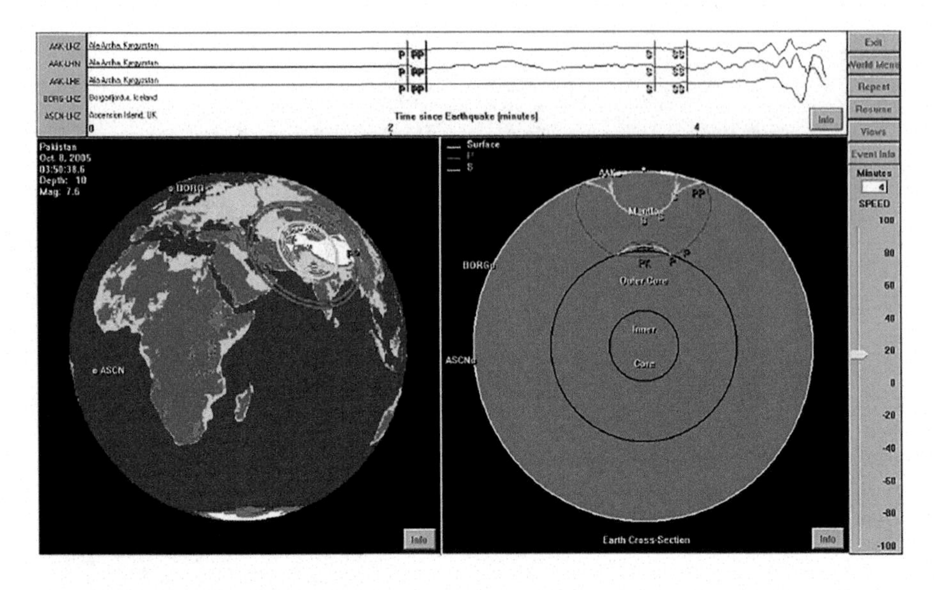

FIGURE 5.13 Pakistan's earthquake analysis through the Alan Jones Seismic Waves Version 2.1. (Alan Jones Seismic Waves Version 2.1.)

local magnitude value that is most often greater than its equivalent counterpart will not be unjustified for using it for CIVWEES for the purpose of optimistic realization between the two at the initial phase. Furthermore, the surface wave magnitude is calculated from the relation $Ms = \log_{10} (amp / T) + 1.66 * \log_{10} (dist) + 3.3$ [16,17], where T is the period of the wave used in seconds, amplitude is in micrometers (this is derived from the min/max A/D counts and the magnitude correction factor) and distance is in degrees.

5.6 WIRELESS TECHNOLOGY COMPARISON FOR CIVIONICS DEVELOPMENT

5.6.1 Telecommunication Infrastructure

The communication infrastructure for Civionics is being designed to characterize the system to meet the multihazard needs of the country. For CIVEEWS, different options are being demonstrated for COS connected with CDPS and the latter with CCCC. The best option is PTCL's (Pakistan Telecommunication Limited) tentative-wired optical DSL broadband network but its existence at many areas at risk is lacking.

In view of such a situation, one has to look for a wireless-cum-wired solution [18] wherein COS is connected with CDPS either through WiMAX (wireless interoperability for microwave access) IEEE 802.16e/m protocol, or Wi-Fi, or Zigbee or through point-to-point radio wave communication technologies or via satellite communication channels or all for ensuring a link redundancy parameter. Similar options have been demonstrated for establishing reliable, fast, robust and secure

TABLE 5.2
Comparative Analysis of Wireless Technologies for Civionics Development

Technology	Wi-Fi	Zigbee	WiMAX
Application	Wireless LAN, Internet	Sensor networks	Metro area broadband Internet
Family	802.11 a/b/g/n	802.15-4, 802.15-4a	802.16 e/m
Typical range	100 m	70–300 m, 900 m	Up to 50 km
Frequency range	2.4 GHz, 5.8 GHz	2.4 GHz, 868 GHz, 915 MHz	2–11 GHz
Data rate	108–600 Mbps	250 Kbps	75 Mbps
Modulation	DSSS	DSSS	QAM
Network	IP & P2P	Mesh	IP
Access protocol	CSMA/CA	CSMA/CA	Request/grant
Key attributes	Wider bandwidth, flexibility	Cost, power	Throughput, coverage
Network topology	Infrastructure/Ad hoc	Ad hoc	Infrastructure
Spectrum type	ISM unlicensed/licensed	ISM unlicensed/ licensed	Licensed
Radio technology	OFDM/MIMO	–	MIMO/SOFDMA
Battery life (Days)	5	1000	3

FIGURE 5.14 Screenshots of developed splendid SMS facility for Civionics. (Develop on Microsoft. NET framework Version 2.0 or above running in Windows 98/2000/XP/2003/ Vista.)

communication between CDPS, CCCC and other components of Civionics such as an Flood Early Warning System, Tsunami Early Warning System, Terror Bombing Early Warning System, and so forth. A comparative analysis of the wireless technologies considered and used for Civionics design is given in Table 5.2. For analyzing the communication link quality and time windowing operations of the various early warning systems of Civionics, we have simulated several scenarios at Hamdard University Communication and Engineering Research Lab [19] on OPNET Modeler Version 15.0/16.0 research and development software.

Hamdard University Pakistan [20] under OPNET's university program [21] has signed a licence agreement with M/S OPNET Incl. in the United States, for using OPNET R&D Modeler purely for research purposes. Similarly, the Satellite Toolkit (STK) [22] basic version has also been received from M/S AGI USA in totting up OPNET with STK for obtaining the simulated results of Civionics while considering the satellite link option. Various simulation results are being published whereby effort has been made to minimize the time window up to 2–120 s for CIVEEWS and also to ensure communication link quality, throughput characteristics, jitter, latency and packet loss for other disasters using WiMAX, Wi-Fi and/or Zigbee models based on IEEE 802.16e/m, IEEE 802.11a/b/g/n and IEEE 802.15-4(a) standards, respectively.

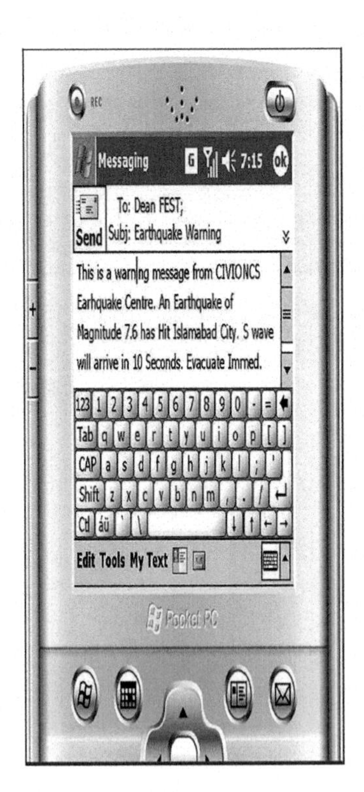

FIGURE 5.15 Text-based warning message.

FIGURE 5.16 Multimedia-based warning message.

5.7 DISSEMINATION OF WARNING MESSAGES

The warning messages after the occurrence of the disaster event are being transmitted to the people at risk, first responders, hospitals, installations at risk, media, and so on, automatically [23]. For the Civionics system, we have developed Splendid SMS delivery software over Microsoft.NET Framework Version 2.0 or above running in the Windows 98/2000/XP/2003/Vista environment. Screenshots of the software are shown in Figure 5.14. The alert messages are delivered as soon as the occurrence of the disaster event is detected by the installed sensors so as to save the lives and property of the people and to give a rescue call to the first responders. Efforts are underway to leverage the concept of femtocell networks [14] possessing capacity gains and ubiquitous indoor coverage bearing the strength to complement macro-cellular networks in spreading warning messages. Figures 5.15 and 5.16 depict the message format being an SMS alert for an artificial seismic activity. These types of messages can be sent to the first responders and the public under disaster during the occurrence of an emergency situation.

SUMMARY

In this chapter, the architecture of an earthquake early warning system is developed and simulated using OPNET telecommunication software for a developing country as a use case scenario. The seismic zones are identified and data processing stations are installed in a simulated environment. The aim is to generate timely response alerts in the case of seismic activity in a country.

REFERENCES

1. Lindholm, C., Hilmar Bungum, Z.R., Lang, D., "*Seismic Hazard Analysis and Zonation for Pakistan, Azad Jammu and Kashmir*," Pakistan Meteorological Department and NORSAR Norway, Islamabad, 2007.
2. International Seismological Center. Available at http://www.isc.ac.uk, [Last accessed on April 15, 2010].
3. United States Geological Survey. Available at http://earthquake.usgs.gov/
4. Pakistan Meteorological Department. Available at www.pmd.org.pk, [Last accessed on April 11, 2010].
5. Incorporated Research Institutions of Seismology (IRIS). Available at http://www.iris.edu/hq/audience/researchers, [Last accessed on January 19, 2010].
6. Lee, W.H.K., "*Earthquake Early-Warning Systems: Current Status and Perspectives*," United States Geological Survey, Menlo Park, CA, 94025, USA. Available at http://www.cires.org.mx/docs_info/CIRES_007.pdf
7. Colibrys Sensor Technology. Available at http://www.colibrys.com/files/pdf/products/DS%20SF3000L%2030S.SF3000L.D.03.09.Pdf
8. Environmental Systems & Services Pvt. Ltd, PO Box 939, Hawthorn, VIC 3122, Australia. Available at www.esands.com and http://www.silicondesigns.com/pdfs/1221.pdf
9. Website Güralp Systems. Available at http://www.güralp.com/products/ [Last accessed on April 14, 2010].
10. Website Güralp Systems. Available at http://www.güralp.com/articles/20060308-casestudy-railway/support, [Last accessed on January 19, 2010].
11. Ventura, C.E., Earthquake Research Laboratories, University of British Columbia. Available at https://windupradio.com/earthquakes/
12. Aldo Zollo, D.C.S., "Real Time Location for a Seismic Alert Management System-Development, HW/SW Integration, Definition and Study of Velocity Models," PhD Thesis, Universit`a di Bologna: Campania Region, Southern Italy, 2008.
13. Wen, K.-L. et al., "Earthquake Early Warning Technology Progress in Taiwan," *Journal of Disaster Research*, Vol. 4, No. 4, 2009.
14. Mutafungwa, E., Hamalainen, J., "Leveraging Femtocells for Dissemination of Early Warning Messages," *Proceedings of IEEE Communications Workshop 2009 Dresden*, June 14–18, 2009, pp. 1–5, Print ISBN: 978-1-4244-3437-4.
15. Wu, Y.-M., Kanamori, H., "Development of an Earthquake Early Warning System Using Real-Time Strong Motion Signals" *Sensors*, Vol. 8, 2008: pp. 1–9. Available at www.mdpi.org/sensors [Last accessed April 1, 2010].
16. Cochrani, L. (24 Garden St, Redwood City, CA 94063). Available at http://www.seismicnet.com [Last accessed on January 19, 2010].
17. Endwell NY 13760 software Seismic Waves Version 2.1 level 2009.9.27.
18. Chris Oberg, J., Whitt, A.G., Mills, R.M., "Disasters Will Happen—Are You Ready?" *IEEE Communications Magazine*, January 2011: pp. 36–42.

19. Disaster Resource Network, "Emergency information and communications technology in disaster response, final report." 2007. Available at http://www.drnglobal.org/index.html [Last accessed on December 15, 2010].
20. Hamdard University Karachi. Pakistan website. Available at www.hamdard.edu.pk [Last accessed on January 19, 2010].
21. M/S OPNET Inclusion USA. Available at www.opnet.com [Last accessed on January 19, 2010].
22. M/S Analytical Graphics. Available at www.agi.com [Last accessed on January 19, 2010].
23. Mase, K., "How to Deliver Your Message from/to a Disaster Area?" *IEEE Communications Magazine*, January 2011: pp. 52–57.
24. Westermo et al., 1980. home.iitk.ac.in/~vinaykg/Iset428.pdf
25. Cua, G., Heaton, T., "The Virtual Seismologist (VS) Method: A Bayesian Approach to Earthquake Early Warning." In: Gasparini, P., Manfredi, G., Zschau, J. (eds.) *Earthquake Early Warning Systems*, 2007: pp. 97–132 Berlin, Heidelberg: Springer.
26. Bose, M., Wenzel, F., Erdik, M., "PreSEIS: A Neural Network-Based Approach to Earthquake Early Warning for Finite Faults." *Bulletin of the Seismological Society of America*, February 2008: 98(1).

6 Modelling and Simulation of a Civionics Multihazard Early Warning System

INTRODUCTION

While designing the proposed early warning system for Pakistan, every effort was made to accommodate a multihazard approach for Civionics. The hazards considered are earthquakes, tsunamis, floods, fires and blasts. The initial communication infrastructure has been suggested for the cities as depicted in Figure 5.4 of Chapter 5. However, the components of such a system can be extended to more regions according to public need and the nature of the disaster once the basic setup is in place. The proposed architecture is wireless cum wired.

6.1 MODELING AND SIMULATION OF A CIVIONICS MULTIHAZARD EARLY WARNING SYSTEM

The following submodels were designed depending on the users' needs:

1. Civionics Wi-Fi cum WiMAX model incorporating random way mobility
2. Civionics Zigbee model with and without mobility
3. Civionics satellite model with multiple orbits

These models were designed and their simulation carried out using the OPNET Modeler Version 14.5 [1] and AGI's STK (Satellite Tool Kit) software [2]. Results were compiled over QoS performance matrices of link quality, packet status, jitter, and so on. The Civionics Wi-Fi cum WiMAX model is shown from Figures 6.1 through 6.6. WiMAX holds the cost advantage in remote areas where wire-line networks are difficult to install and are an expensive alternative. In our model, the selection of traffic was made to accommodate packets of data emanating from the Earthworm software on triggering of the disaster event (in case of earthquakes). Other traffic loads such as web browsing, email and multimedia information transmission have also been selected in profile and application definitions of the OPNET models [1]. The subnets located in the regions or cities are connected through routers, switches, hubs and WiMAX base station antennas using ATM/Sonnet links of varying capacities. During the generic design phase of the Civionics satellite model, multiple orbits were tried using the STK orbit directory. Ikonos and GPS 2-11 [2] have been found to have a footprint for Pakistan's geographical region. For capitals like Karachi, the design includes mobility models as well.

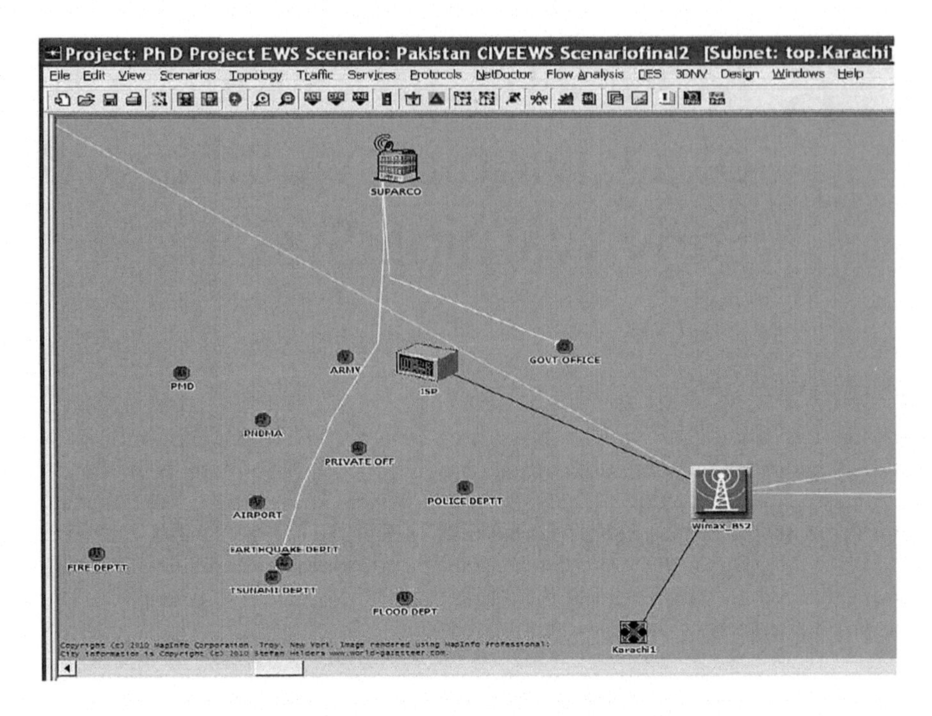

FIGURE 6.1 OPNET model—interconnecting departments through ISP.

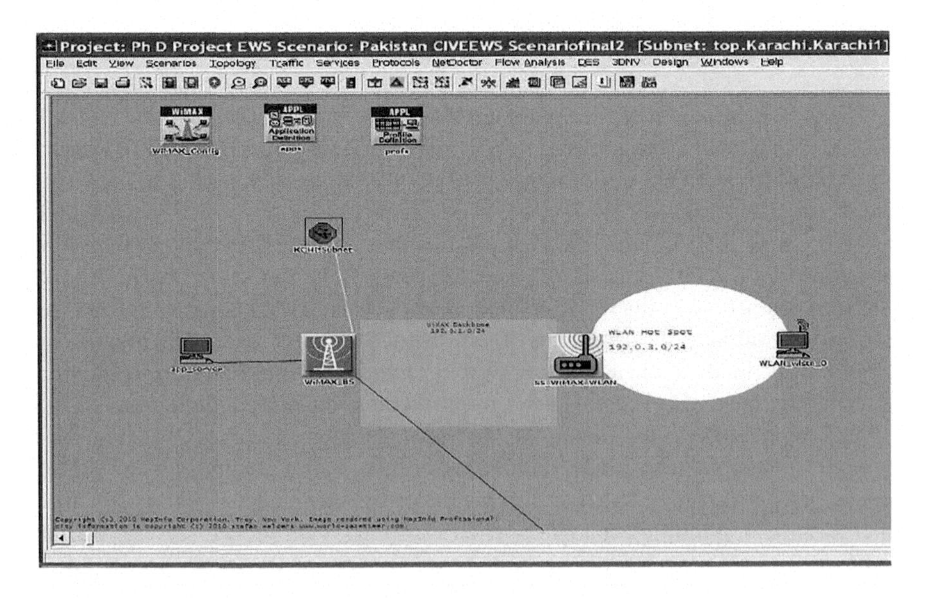

FIGURE 6.2 OPNET model—one of the WiMAX stations in Karachi.

FIGURE 6.3 OPNET model—one of the stations connected through satellite.

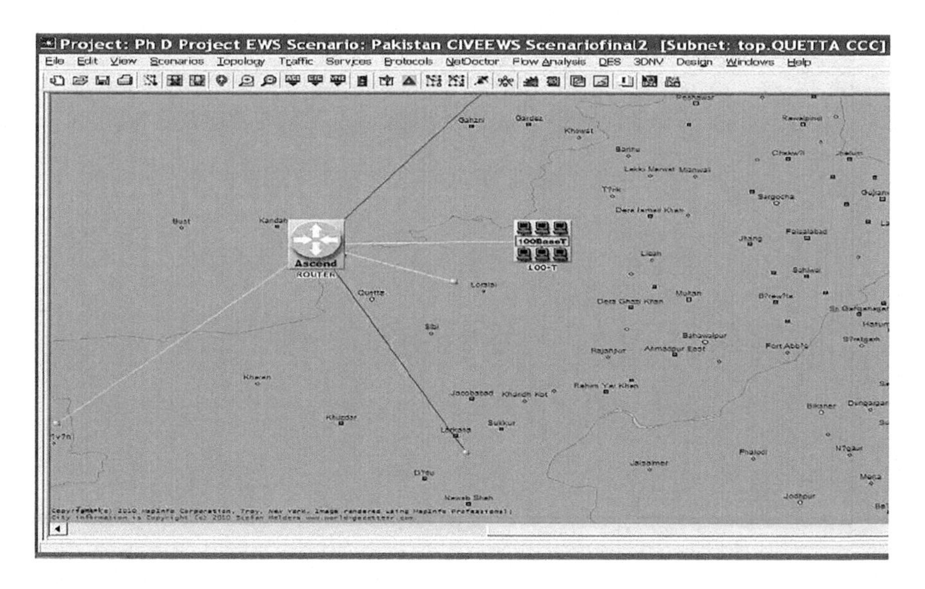

FIGURE 6.4 OPNET model—100 base T station connectivity and router at Quetta.

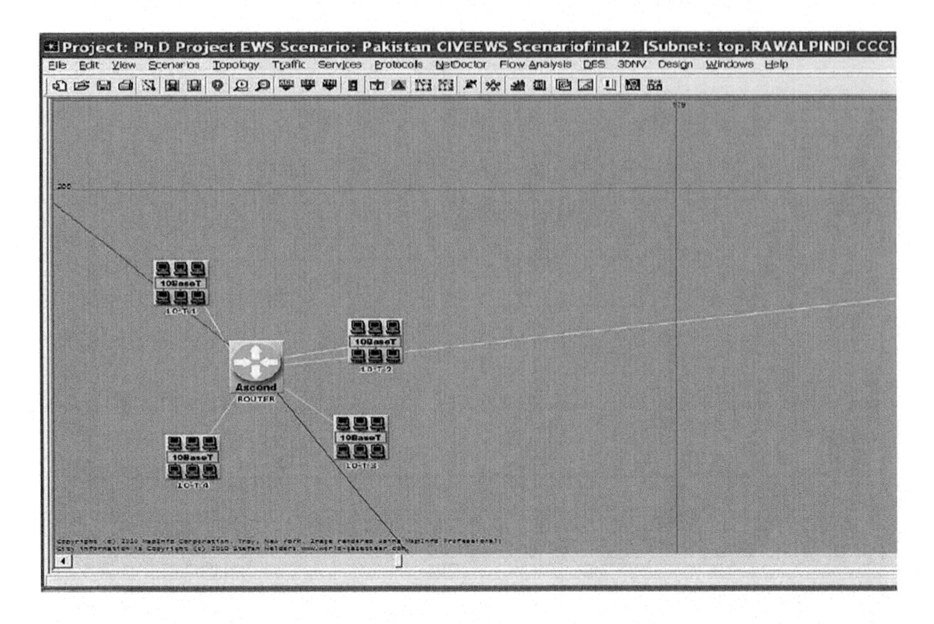

FIGURE 6.5 OPNET model—10 base T station connectivity with router at Quetta Command and Control Centre.

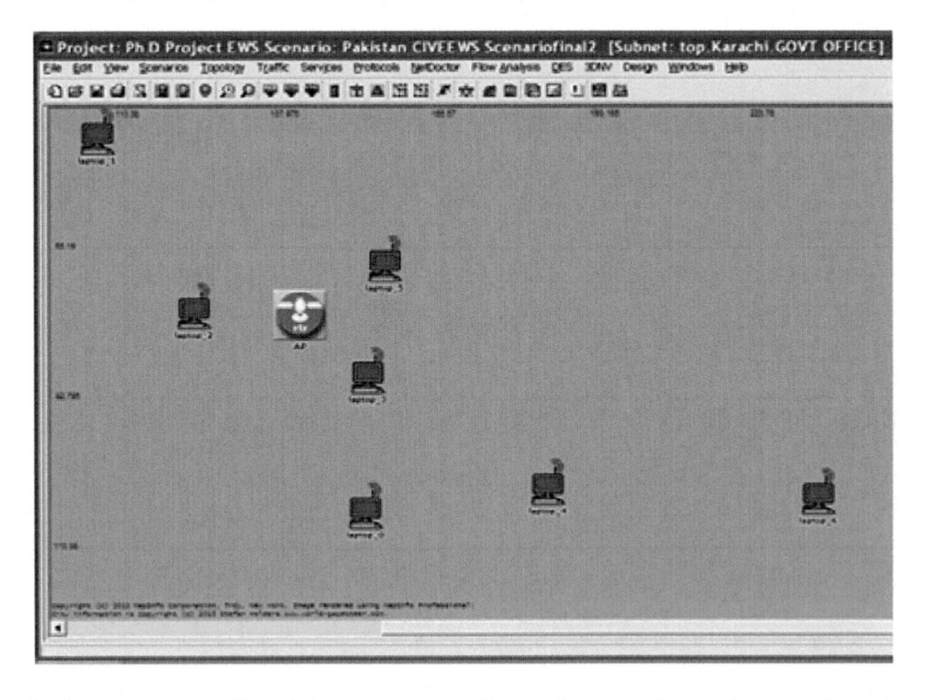

FIGURE 6.6 OPNET model—Government office Wi-Fi connectivity of laptop nodes with other networks.

TABLE 6.1
Inventory Record of the OPNET Scenario

S. No.	Element	Type	Count
1	Devices	Total	212
2		Routers	60
3		Switches	3
4		Hubs	1
5		LAN Nodes	11
6		Workstations	116
7		Servers	1
8	Vendors	Alcatel-Lucent	5
9	Physical links	Total	100
10		ATM	28
11		Serial	36
12		Ethernet	16
13	Other	Configuration Utilities	64

Figure 6.1 suggests intercommunication and information dissemination regarding the disaster event among varying agencies and departments through an ISP located in each of the provincial headquarters.

Table 6.1 shows a record of OPNET's inventory used in designing the EEWS model.

6.1.1 Simulation Graphs

Figures 6.7 through 6.14 show a few simulation results of the Civionics multihazard early warning system.

6.2 DESIGNING THE CIVIONICS EARTHQUAKE EARLY WARNING SYSTEM—ZIGBEE MODEL

The following section defines and depicts the Civionics Zigbee model. Zigbee is the communication standard of IEEE 802.15-4. These tiny, inexpensive and easy to install devices incorporate smart sensor technology which triggers on the occurrence of a disaster event. They bear the models using transceiver characteristics. A network of 65,000 nodes is permissible, providing a range of ~900 m. The proposed model as shown in Figure 6.15 suggests a communication network of seismic-prone regions or tentative disaster sites using the Zigbee network of nodes. These nodes have a communication capability with Zigbee routers and coordinators. The model has also been accommodated with satellite connectivity among various subnets using Ikonos and GPS orbits of STK so as to allow link redundancy characteristics of Civionics. The OPNET model Zigbee nodes are shown at Karachi subnet in Figure 6.16 and the OPNET model Zigbee nodes are shown for Rawalpindi CCC in Figure 6.17; however, Figure 6.18 depicts the WiMAX scenario at a satellite subnet.

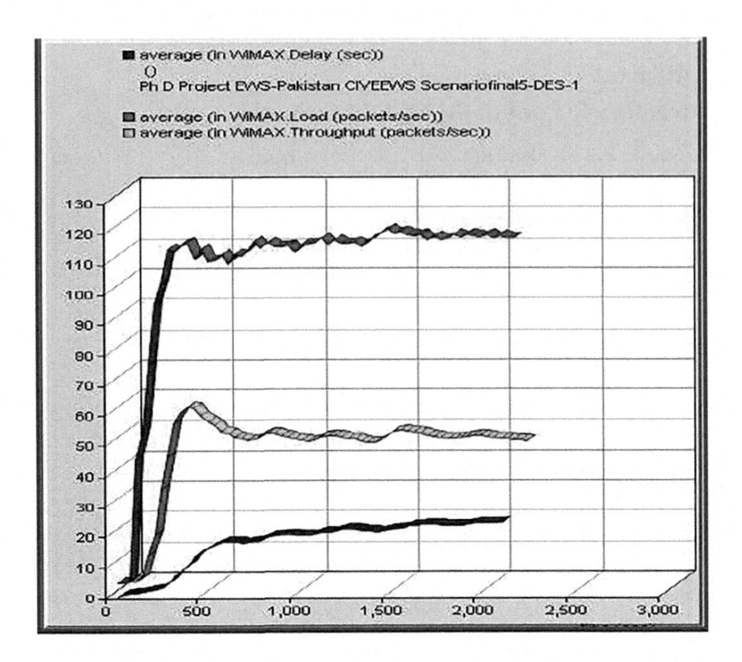

FIGURE 6.7 Delay v/s load v/s throughput at WiMAX nodes (packets/s).

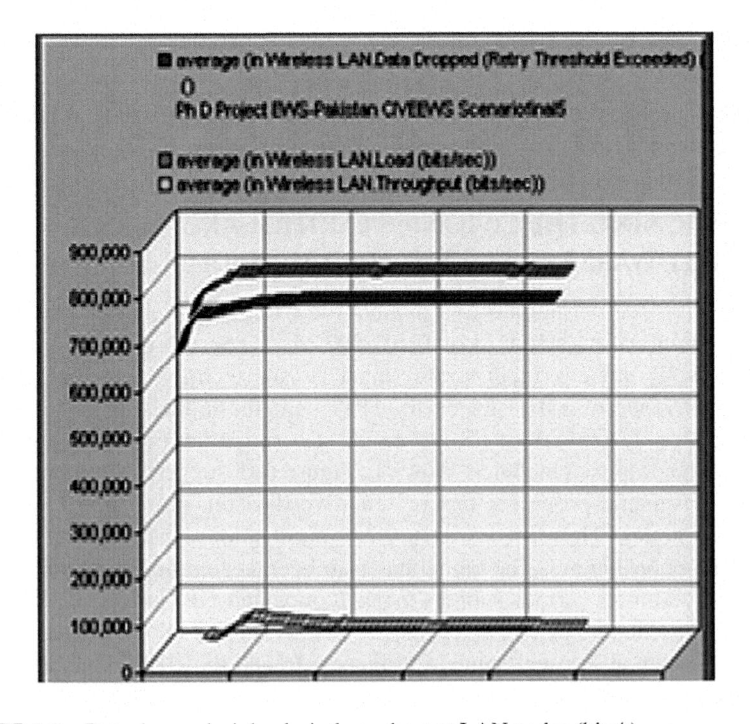

FIGURE 6.8 Data dropped v/s load v/s throughput at LAN nodes (bits/s).

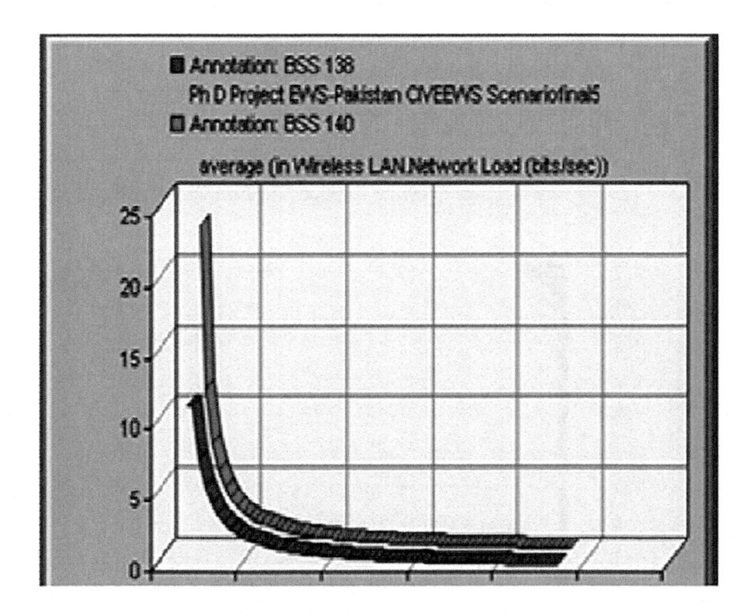

FIGURE 6.9 Annotation at BSS 138 v/s BSS 140 (bits/s).

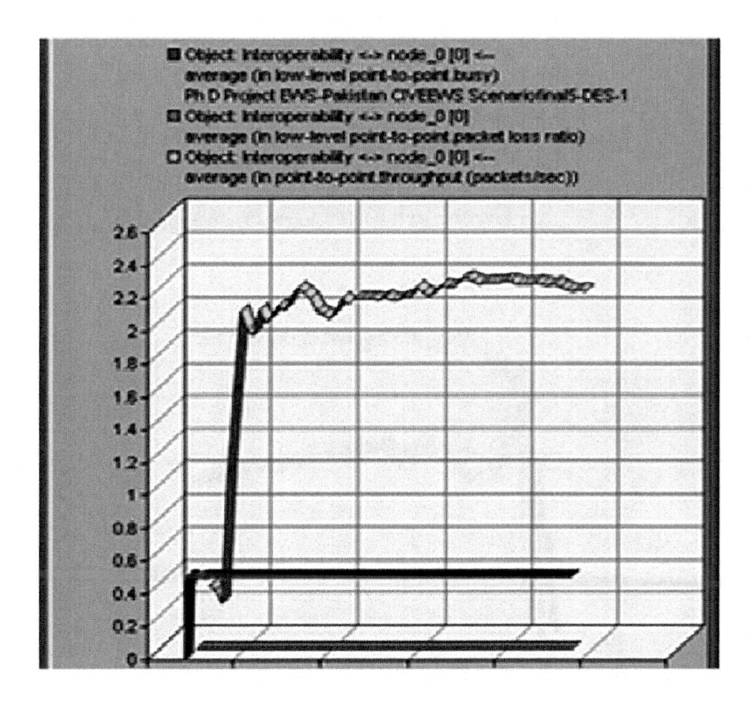

FIGURE 6.10 Traffic at subnet interoperability—busy v/s packet loss v/s throughput.

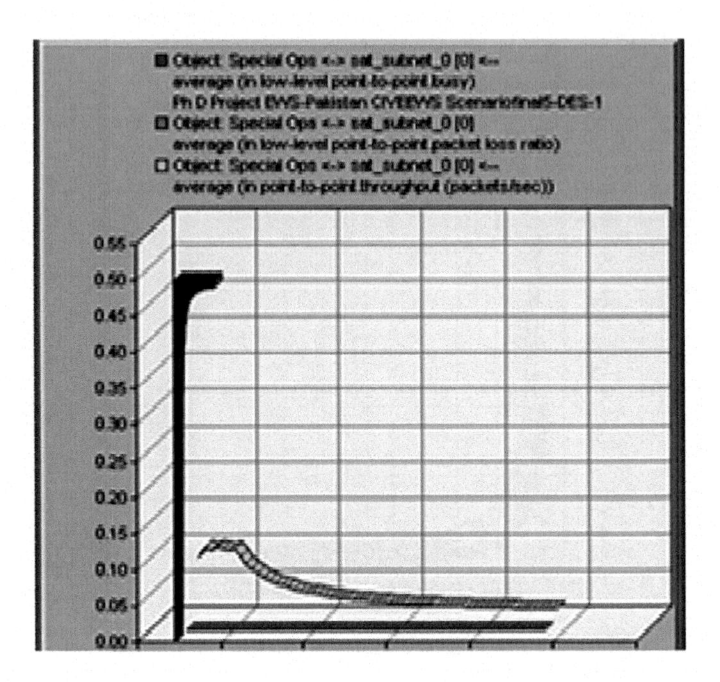

FIGURE 6.11 Traffic at satellite subnet—busy v/s packet loss v/s throughput.

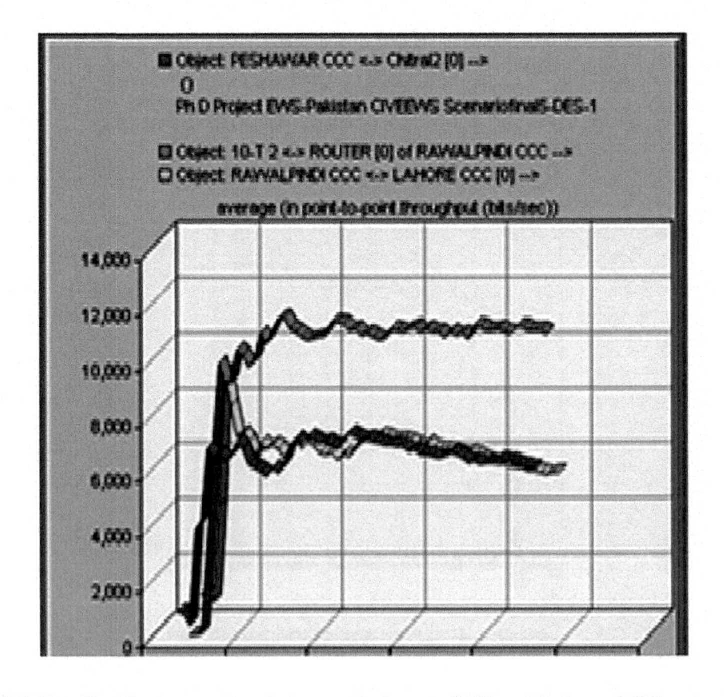

FIGURE 6.12 Traffic comparison between Peshawar CCC and Lahore CCC.

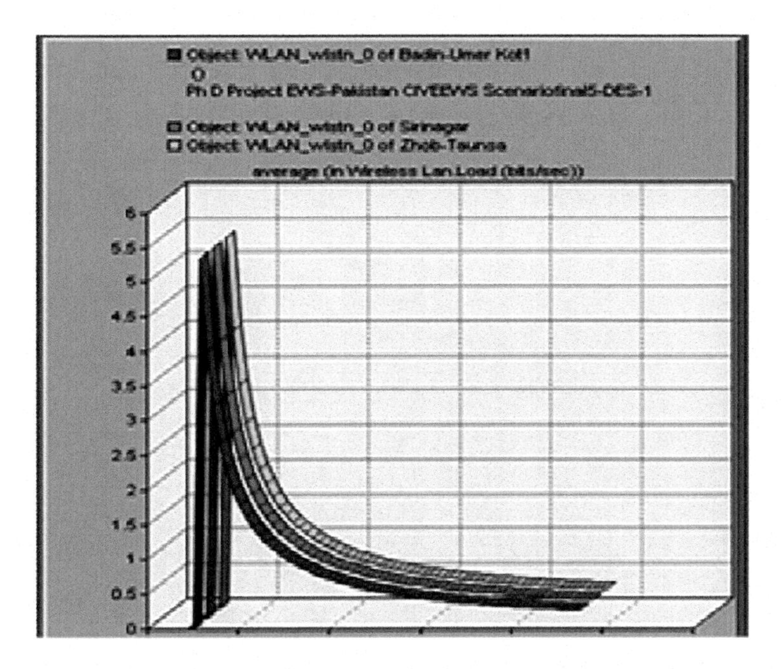

FIGURE 6.13 Wireless traffic comparison between subnets Badin, Srinagar and Zhob.

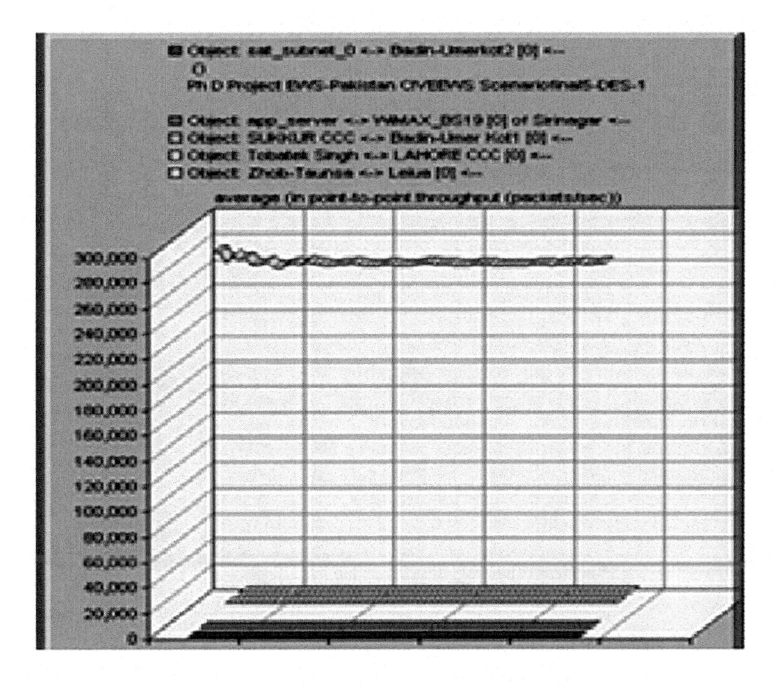

FIGURE 6.14 WiMAX traffic at BS-19 application server and other subnets.

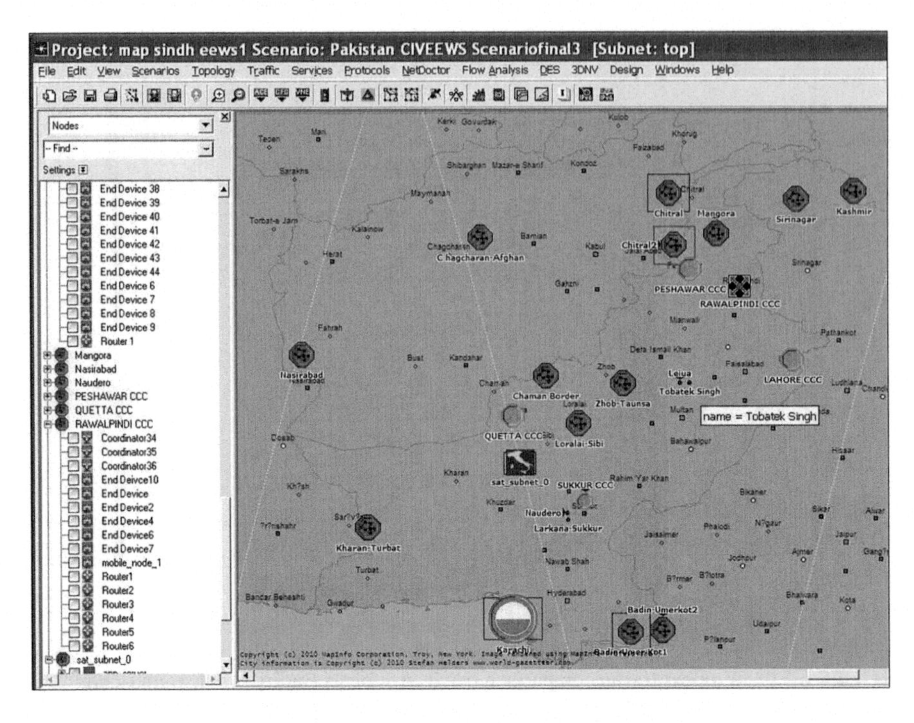

FIGURE 6.15 OPNET model—CIVEEWS Zigbee and satellite model.

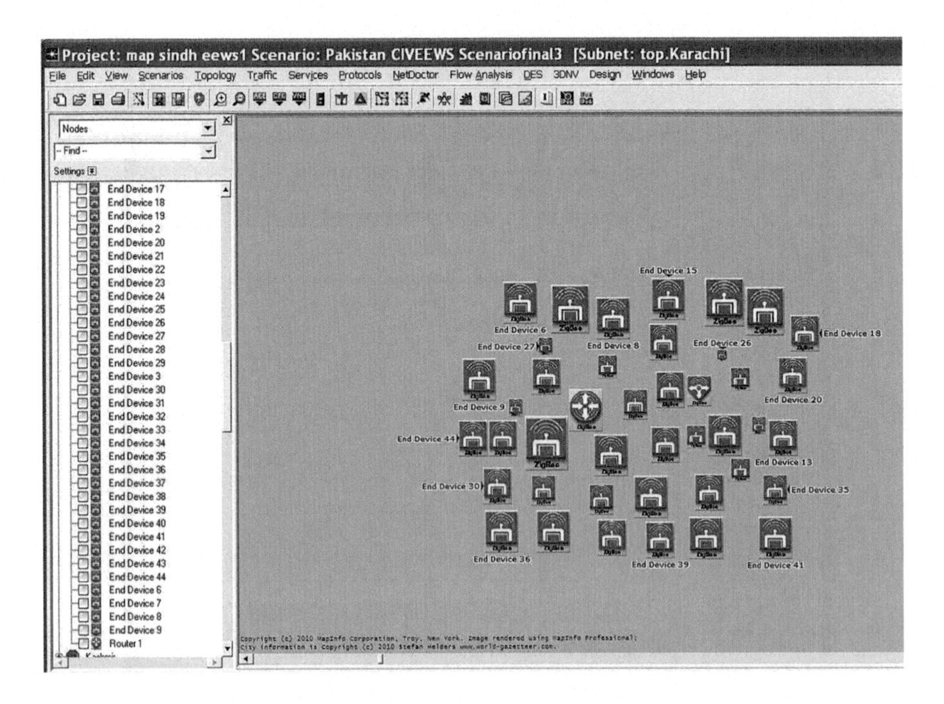

FIGURE 6.16 OPNET model Zigbee nodes shown at Karachi subnet.

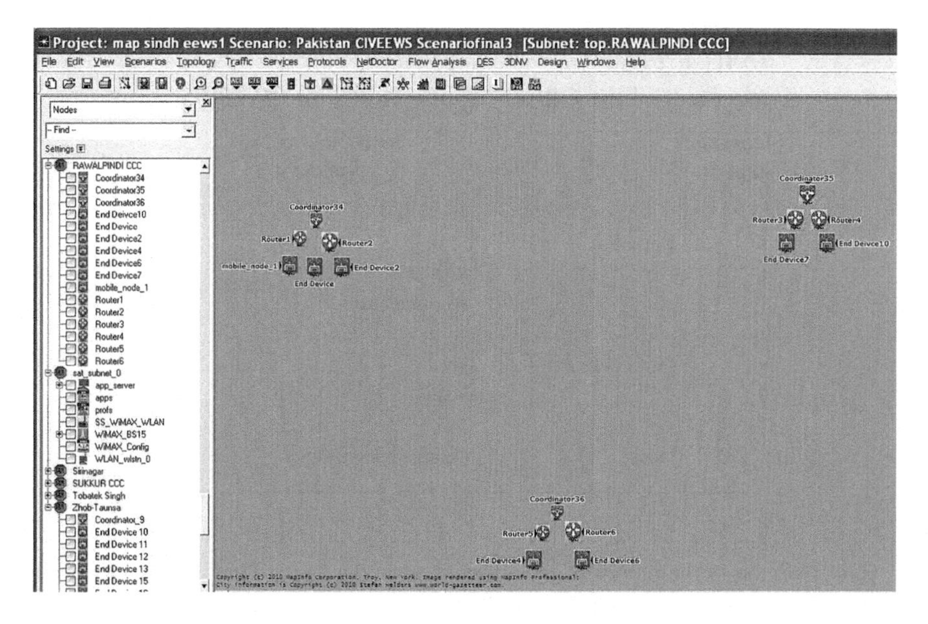

FIGURE 6.17 OPNET model Zigbee nodes shown for Rawalpindi CCC.

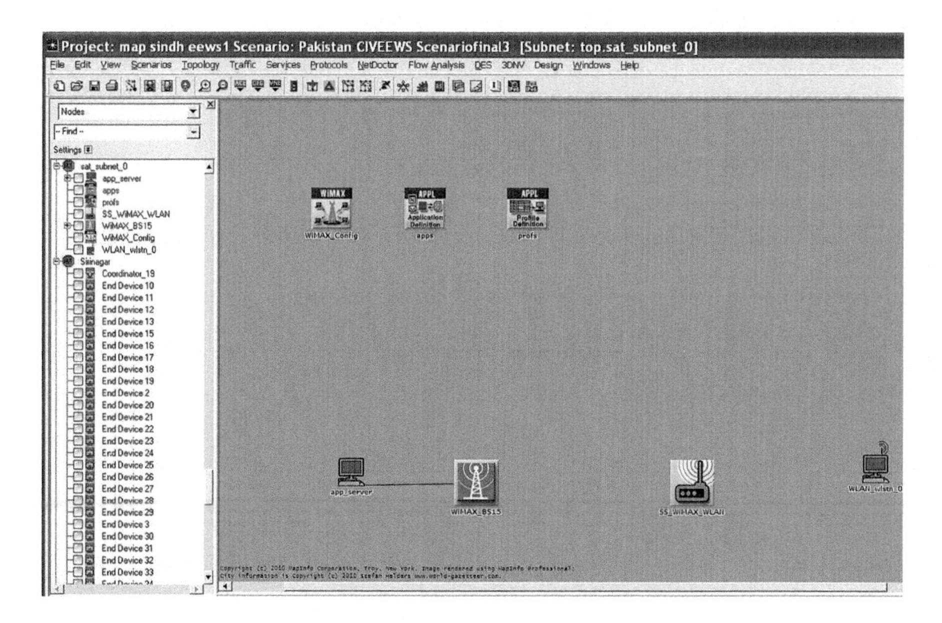

FIGURE 6.18 WiMAX scenario at the satellite subnet.

TABLE 6.2
Inventory of the OPNET Zigbee Model for CIVEEWS

Category:		**Network Inventory**	
Report:		**Summary**	
		Total	798
1	Devices		
2		Routers	2
3		Workstations	2
4			
5	Physical Links	Total	1
6		Serial	1
7			
8	Other	Configuration Utilities	4

Table 6.2 shows the record of inventory used in designing the OPNET model of Civionics early warning scenario with Zigbee.

6.3 SIMULATION GRAPHS

Figure 6.19 shows the simulation progress of the scenario, whereas the following are graphs showing a few results of the scenario of Figure 6.15.

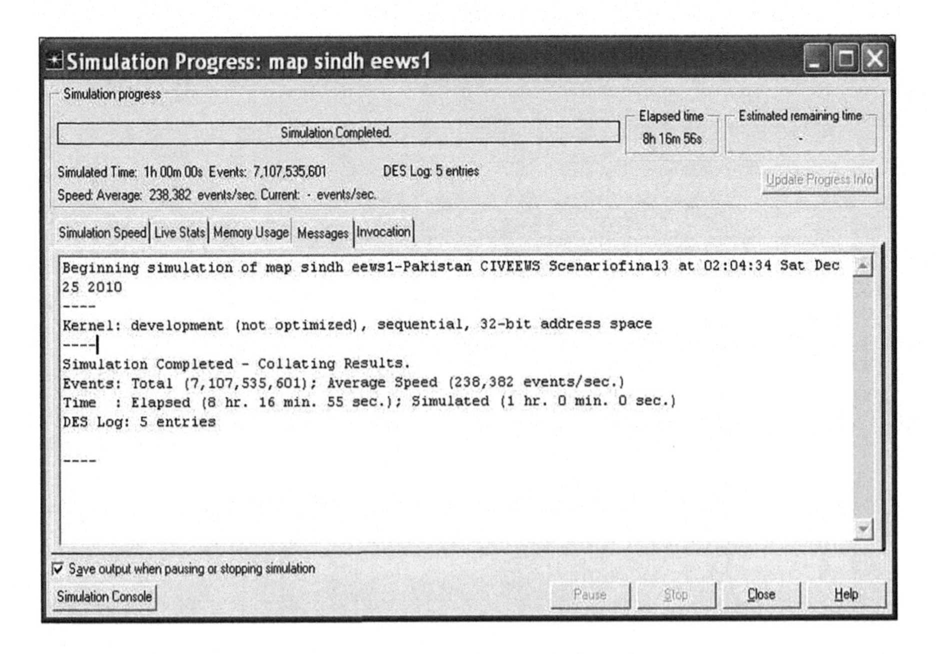

FIGURE 6.19 Simulation progress in OPNET for EEWS.

6.3.1 Results and Discussion

The objective of this chapter is to leverage the abilities of available seismic sensors to develop a real-time multihazard early warning system that is being installed using realistic simulation results achieved through the OPNET Modeler to generate robust, timely and automatic warning messages and alerts for the communities and installations at risk. To achieve these objectives, a few data sets were collected and analyzed for observance initially of the design limitations of Civionics EEWS based on the set criteria that the local magnitude is greater than 5.0, focal depth <40 km and the earthquake occurred within 50 km of the identified zones. After processing the data sets with Winquake and the individual waveforms interfacing with the OPNET Modeler, the CIVEEWS network architecture is analyzed with varying communication models such as Wi-Fi, WiMAX, Zigbee and others over the performance matrix of link quality, link reliability, guaranteed data transfer, error control, QoS, jitter, packet loss and throughput for fixing up the minimum warning time window. Several results were obtained for varying models and a few are shown in Figures 6.20 through 6.25. A careful observation of the results follow that wireless networks such as WiMAX, Wi-Fi and Zigbee are competitive technology solution for early warning systems in multihazard scenarios. To address the issue of minimizing the timing window for early warning messages in natural disasters such as earthquakes, the research community has to rely on the networks with minimum latency, fast transmission rate, less jitter, guaranteed QoS and reliable and secure data transfer. Congestion in networks used in emergency situations requires that a femtocell networking perception be developed among the disaster management communities. This chapter also discusses the architecture for a

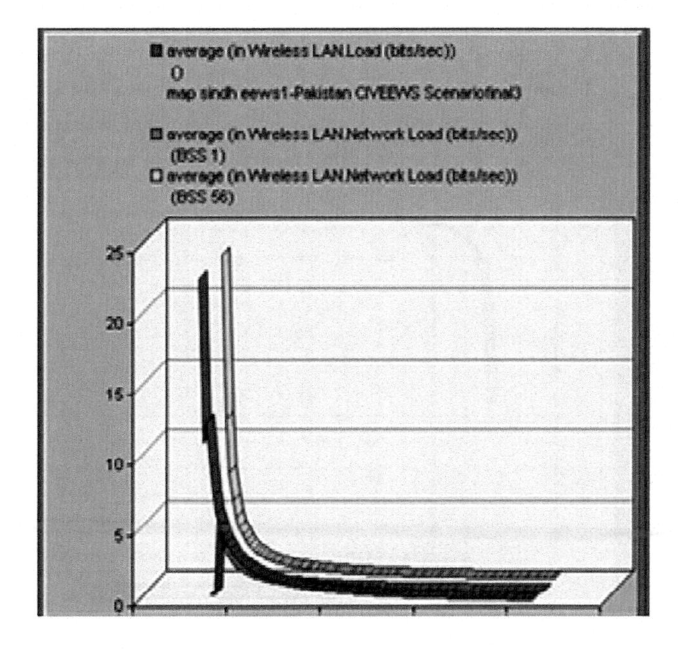

FIGURE 6.20 Wireless LAN network load (bits/s).

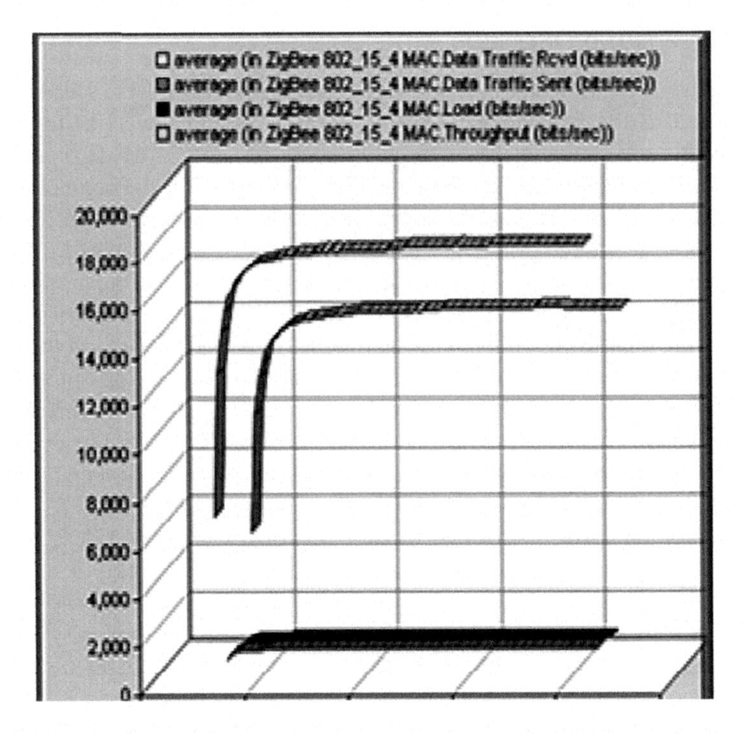

FIGURE 6.21 Traffic received v/s sent v/s throughput at Zigbee nodes.

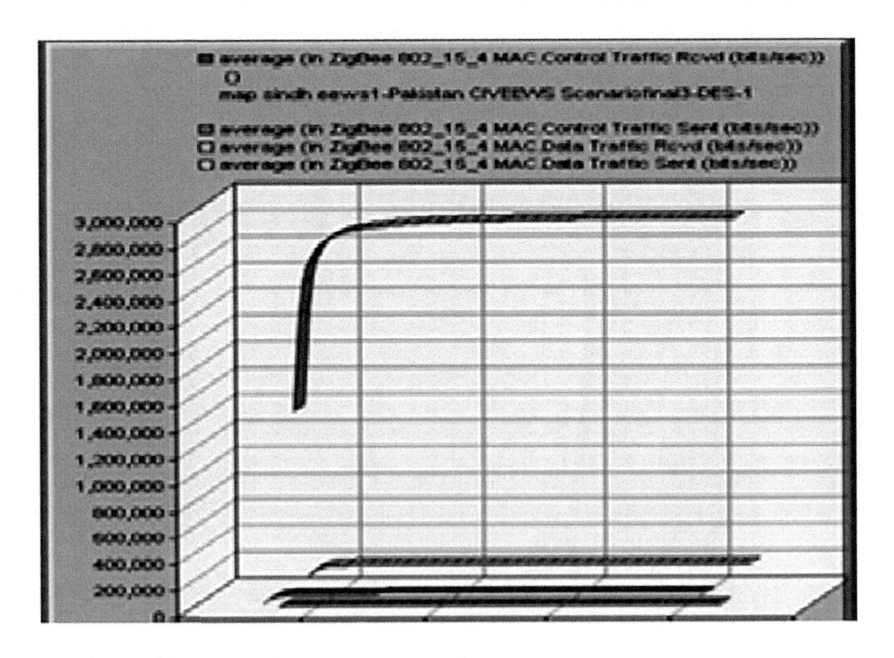

FIGURE 6.22 Control traffic received v/s sent (bits/s) at Zigbee nodes.

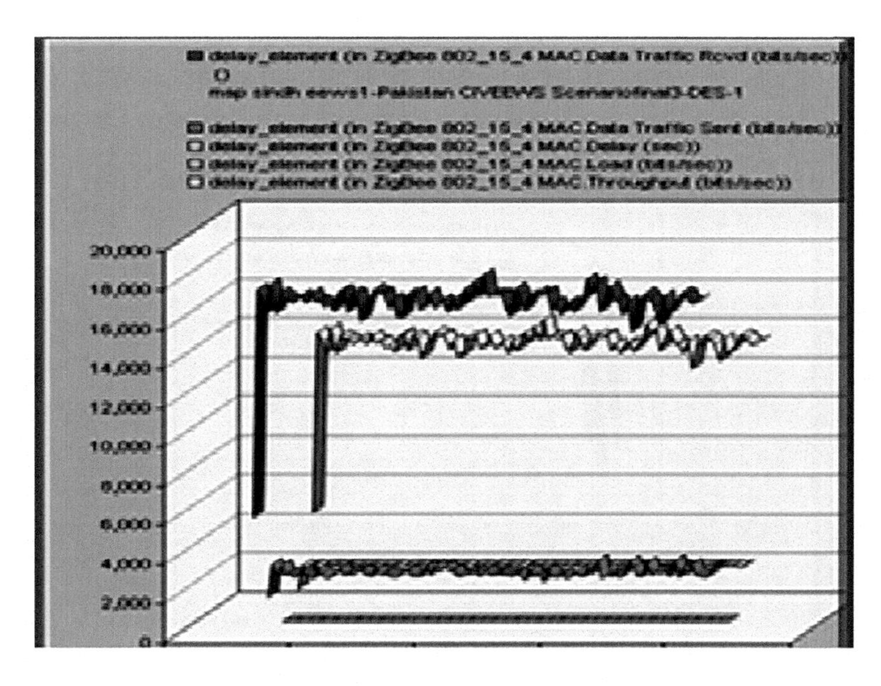

FIGURE 6.23 Data traffic received v/s sent v/s throughput at Zigbee nodes.

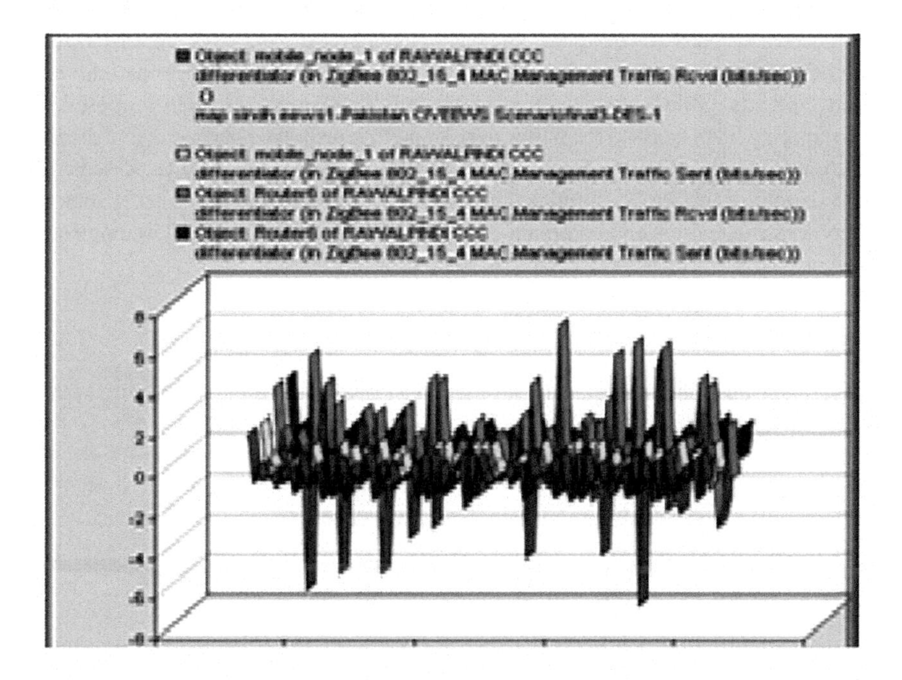

FIGURE 6.24 Traffic received v/s sent at Rawalpindi CCC (bits/s).

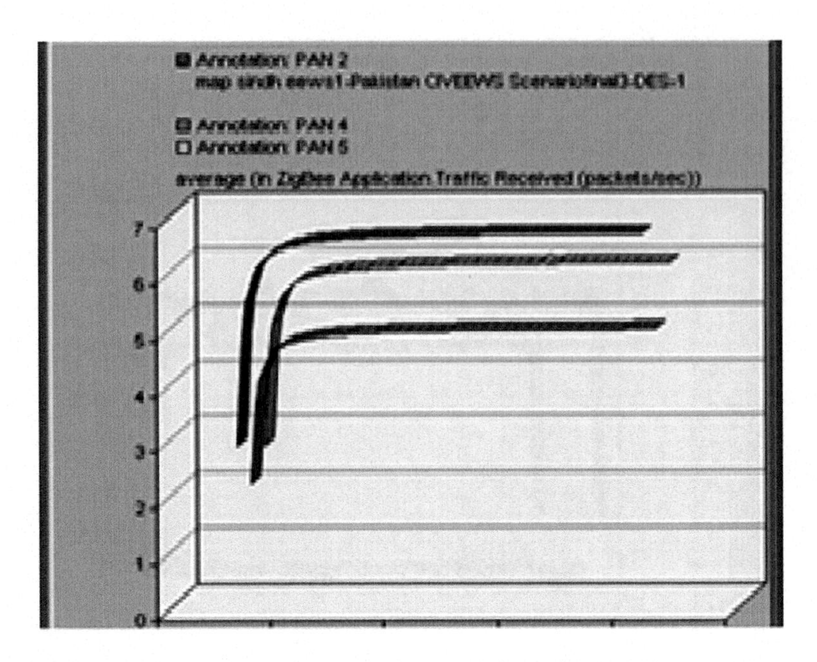

FIGURE 6.25 Annotation comparison between PAN 2, 4, 5 (packets/s).

multihazard early warning system with emphasis specifically on the earthquake early warning system called CIVEEWS. The results of a minimized warning time window for stations at distances of 15–27 km, using various simulated communication models developed in OPNET for design of an earthquake early warning system for Pakistan called CIVEEWS, are given. Apart from this, a rapid warning dissemination setup called "splendid SMS" has also been proposed. This setup has been proposed for disseminating SMS alerts to the public at risk and first responders ahead of any disaster The disaster engineering community can also benefit from the novel idea presented in this research for invoking a simulated technique for fixing up on-site seismic stations, data processing centres and command and control centres used in the development of a new or upgrade of existing early warning systems.

SUMMARY

This chapter describes the essentials of a robust multihazard early warning system that responds to all hazards and disasters, such as earthquakes, tsunamis, floods, fires and explosions, and produces timely alerts and warning messages to save lives and property of the affected population. The design is supported by a real-time simulation environment in a use case scenario of a developing country.

REFERENCES

1. M/S OPNET Inclusion USA. Available at www.opnet.com [Last accessed on January 19, 2010].
2. M/S Analytical Graphics. Available at www.agi.com [Last accessed on January 19, 2010].

7 Multihazard Disaster Engineering during the Response Phase

INTRODUCTION

Multihazard early warning systems using Wi-Fi, WiMAX and Zigbee, and satellite communication technologies using wireless cum wired communication networks across the country have been proposed. Once these networks are in place across the country, the disaster handling will be more efficient and easier during the response phase. This chapter discusses the disaster engineering technologies after a disaster has hit any region.

7.1 INFORMATION AND COMMUNICATION TECHNOLOGIES AND NETWORKS FOR DISASTER MANAGEMENT

Talking in a broader sense, any disaster engineering network needs the following technologies to be implemented in its communication networks [1,2]:

- TETRA, TETRA II, and TETRAPOL
- Project 25
- 3G cellular systems
- IEEE 802.11×
- IEEE 802.16×
- IEEE 802.20
- DVB: DVB-T, DVB-S, and DVB-H
- HC-SDMA
- HIPERLAN2
- Interactive satellite broadcasting systems and GPS
- IEEE 802.15×
- UWB
- Sensors
- GIS
- RFID

Two computer-based tools are also necessary, which should be attached to the system for disaster handling—the Hazus loss estimation tool and the Sahana Disaster Management Tool. However, there appears to be a research need to invent a Hazus-like tool for Pakistan's geography. We will discuss the importance and implementation of both of these tools in later sections of this chapter.

The popular satellite vendors accessible to Pakistan are:

- *Broadband Global Area Network (BGAN)*: Provided 40 satellite phones to Pakistan during the earthquake response operations in 2005 in Kashmir
- *Thuraya*
- *Iridium*

7.2 FIRST RESPONDER REQUIREMENTS

First responders are the individuals who respond immediately after a disaster hits. Their operations demand all-round 4Cs (command, control, coordination and communication) and information sharing with other departments, law enforcement agencies, firefighters, ambulances and staff, as well as local and international public safety community and media. Information sharing is not merely just voice but also images, videos and all multimedia data over wired and wireless media. Hence, first responders need interoperable communication technologies, devices and channels that have the capability of ubiquitous computing as authorized for all-round communication. First responders therefore must possess not only interoperable communication networks but also the vehicles equipped with such systems that respond during and after the disasters. They need law enforcement communication devices (LECDs) specially designed for the response phase as per the disaster scenarios. A disaster can fall within one of the following scenarios of systems:

- Personal Area Network (PAN)
- Jurisdiction Area Network (JAN)
- Incident Area Network (IAN)
- Extended Area Network (EAN)

Figure 7.1 shows the wireless standards in consonance with the above-mentioned networks. Hence, the LECD must be designed to support all the four

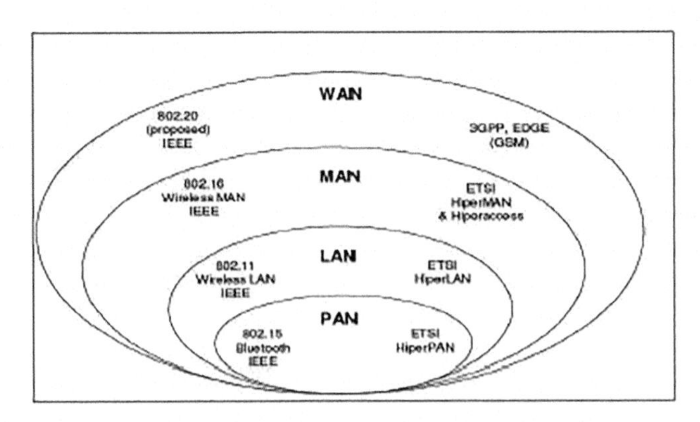

FIGURE 7.1 Wireless standards. (JJECIP for Disaster Response, 2008.)

TABLE 7.1
Mobile Technology Standards

Generation	Standard	Description	Data Rate (kbps)	Throughput (kbps)
2G	GSM	Transfer voice, or low-volume data	9.6	9.6
2.5G	GPRS	Transfer voice, or moderate-volume data	21.4–171.2	48
2.7G	EDGE	Simultaneous transfer voice and data	43.2–345.6	171
3G	UMTS	Simultaneous transfer voice and high-speed data	144–2000	384

aforementioned networks. A comparative analysis of Mobile Telephony standards is shown in Table 7.1.

For ensuring interoperability, the architecture of multiple agencies is the need of the hour. The wireless standards will act as only the interfaces. Such a scheme of interoperability can be established only through routable networks like WiMAX, Wi-Fi and the Internet cloud. The interfaces required for such architecture are LAN interface, SMS, email or data packaging interface, security interface for authentication, QoS interface and management interface.

During the disaster occurrence response phase, the agencies usually experience failure in talking and communicating data to each other. This is due to incompatible equipment, limited and fragmented funding, planning and radio spectrum, lack of coordination and cooperation. The law enforcement agencies may adopt strategies to talk to each other without failure, such as swap radios, talk-around, mutual aid channel, gateway console patch, network roaming and standard-based shared networks.

7.3 4G ACCESS TECHNIQUES FOR DISASTER MANAGEMENT

4G access techniques are the core research area wherein the software defined radio (SDR) approach is being used to interoperate current Wideband-CDMA (WCDMA) with 1 × RIT and 3 × RIT. Large Area Synchronized CDMA (LAS-CDMA) is the approach being worked upon by LinkAir Communications to resolve the issue of interoperability of 3G and 2.5G devices and networks with those of 4G. With 4G, an Internet speed of 100 Mbps can be possible against the present 2 Mbps with 3G. The 4G networks implementation will be in between 2012 and 2015. The target for the 4G systems is an IP, packet-switched network solution where voice, data and streamed multimedia are available "Anytime, Anywhere" for users along with higher data rates than previous generations. The QoS requirement of 4G wireless networks will be on par with the same in wire line environment (99.999% reliability) [3]. In a 4G environment, location-based database access will be possible with the help of Telegeoprocessing—a technology which is a combination of GIS and GPS in fixed or mobile vehicle scenarios. In addition, the law enforcement agencies will be able to undertake Virtual Navigation, Telemedicine and videoconferencing with 4G in disaster engineering's response phase wherein the entire communications infrastructure is destroyed. But the increased cost is one of the limitations of 4G technology.

7.4 COMMUNICATION NEEDS FOR LAW ENFORCEMENT AGENCIES

Law enforcement agencies and public safety organizations need voice and data (including multimedia) communication facilities both within and outside their domains. One has to identify the communicating nodes and their purpose with special demands and constraints such as time limits, encryption and access to critical information. In order to explore the communication equipment, need and country's capability to deal with the same, we have chosen the following three disaster scenarios:

7.4.1 Fires
7.4.2 Blasts
7.4.3 Earthquakes

We will now define the communication needs for these three scenarios.

7.4.1 Fires

The following scenario describes an extreme fire after a blast covering a widespread geographical area. On Monday, December 28, 2009 (day of Ashura, the 10th day of Muharram) at about 4:13 PM, a powerful bomb blasted the Shiite procession near the Light House off Muhammad Ali Jinnah Road, Karachi and the financial hub of the country came to a grinding halt. Initial investigations confirmed that the suicide bomber wearing in his jacket about 20 kg of high explosives entered among the thousands of participants of the procession and blasted him resulting in the fatal suicide attack which caused martyrdoms of more than 100 people while critically injured several others. Some other investigations believe that the bomb was planted in a box with sacred papers that was remotely detonated. In the aftermath of the event, an unrecognized angry mob turned violent and set ablaze around 4000 shops in the neighboring markets after looting them all. More than 50 vehicles, including the fumigating vehicles belonging to the police, rangers and the city district government of Karachi, were also burnt to ashes. This arson resulted in losses of approximately 0.35 billion US$ and deprived more than 10,000 families of their livelihoods. Karachi's big and densely populated markets such as Bolton Market, Light House, Paper Market, Feroze Market, Akber Market. Warehouse in the Plastic Godowns, Kharadar, Mitha Dar Police Sections, Medicine Market, Madina Ice Cream, Denso Hall and Kachi Galli, shops which sell arms and ammunition were also torched through some chemical materials (RDX, TNT, Torpex or H6,TNT = 2,4,6-trinitrotoluene. 2,4,6-Trinitrotoluene is produced commercially by the nitration of toluene and it is used mainly as a high explosive in military and industrial applications. Exposure to 2,4,6-trinitrotoluene both through inhalation and skin absorption can occur during its production, during munitions manufacturing and loading, and during blasting operations. 2,4,6-Trinitrotoluene has been detected in wastewater, surface and groundwater, and in soil and sediment near plants manufacturing 2,4,6-trinitrotoluene and explosives [http://www.inchem.org/documents/iarc/vol65/trinitrotoluene.html], RDX = cyclotrimethylene trinitramine, that catches fire in seconds. The Fire Office officials maintained that they tried to douse the flames with water, but that did not work and thus they had to use foam to extinguish the blaze. The miscreants also fired upon the people, the law enforcement agencies present on the scene, and later the fire brigades when they arrived to extinguish the fires. Some CNG cylinders of the vehicles also got blasted after catching fire. In a separate incident on the same day, the unidentified miscreants set ablaze two buses P-1214 and P-0408 on the National Highway. Three banks, including the Allied Bank, Habib Bank and National Bank of Pakistan, two ATM machines and one traffic signal also came under attack from the miscreants.

4:15 PM

1. A guard calls 115 emergency services that offices, markets and banks situated in Sadar Karachi have been set on fire and office records are liable to be damaged, including the fire threat to the staff present there. The dispatch system notes the complaint in the computer-assisted dispatch system.

4:17

2. The dispatch system notes down the address of the incident, issues a list of units around the area and assigns engines (like E46, E43, E38), battalion commanders, a ladder truck (L43) and an ambulance with a display of job description of each of these.

4:17

3. Fire engines and the battalion commander confirm that they have taken action and are on the way with sirens via their devices. The information of persons in the unit is transmitted automatically.

4:25

4. The computer-assisted dispatch system notifies regarding additional vehicles, fire engines, trucks and rehabilitation vehicles.
5. The system informs all unattended units and officers.

4:30

6. Police vehicles with a sufficient number of law enforcement personnel are notified through the CAD system and those respond by arriving at the scene.

4:35

7. The global positioning system (GPS) guides the fire engines and these confirm the CAD system through communication devices that they have entered at the site of the fire incident.

4:36

8. The commander confirms their presence via the broadcast wireless radio messaging system and provides scenario videos to the CAD system. He acts as the incident commander of the scene.
9. Dispatch orders a power source vehicle and gas source vehicle on request by the CAD system.

4:40

10. The IC assures and announces the exit of everybody on the scene and gets a water source. The ladder truck finds hydrants through digital photographs on its device.

4:45

11. Power and gas resource agencies are now linked into the incident and their availability to the incident is broadcast through devices to the CAD system.
12. The commander can look, hear and guide the engines, emergency vehicles, water resources, power resources and law enforcement personnel. He can communicate with them the voice, data, imagery, maps and resources detail through devices that are interoperable.
13. The commander gives instructions for actions to kill fumes, control fire spread, evacuate public if any, etc.

4:45

14. Engine notifies the CAD system and to the commander about the shortage of water and reduced air pressure.
15. The CAD system notifies the water department about the shortage. The water department is now linked to the CAD system and commander devices. It issues notices to the emergency units which respond on the pager. After some time, water is made available in abundance through fire brigades.
16. The fire is so extreme that the commander calls on all the fire brigades working in the city to respond without fail at the site of the incident.
17. The commander orders aerial cover/help with sufficient staff armed with plastic bullets.
18. Law enforcement agencies order gas shelling, water shelling/shedding and plastic bullets (ground and aerial) to scatter the agitating mob so as to stop them from spreading the fire.

4:50

19. The mob disappears within minutes and the area is completely empty of miscreants' attacks.

11:55

20. The commander senses that the fire is completely extinguished and the operation should now be stopped. He broadcasts evacuation messages to all deployed units and hands over the building's custody to the concerned officials.

7.4.2 Blasts

1. A dump truck filled with explosives detonated in front of the Marriott Hotel in the Pakistani capital, Islamabad on September 20, 2008 at 6:00 PM, killing 58, injuring at least 266 and leaving a 60 ft (20 m) wide, 20 ft (6 m) deep crater outside the hotel. The explosion destroys several vehicles and caused structured damage to dozens of nearby houses. Foreign nationals were also included in the dead and injured. The blast caused a natural gas leak that set the top floor of the 5-storey, 258-room hotel on fire. The massive explosion was heard 15 km away. The blaze that followed quickly engulfed the entire structure of the hotel. About two-thirds of the building caught fire as a result of the explosion after a natural gas pipe was blown open. The truck carrying the bomb was stopped at the front barrier. Some shots were

fired; one of the guards fired back, and in the meantime the suicide attacker detonated all the explosives. All six guards at the gate died. A government spokesman confirmed that approximately 600 kg (1300 lb) of RDX mixed with TNT (Torpex or H6) and a mixture of mortar and ammunition to increase the explosive capacity were used in the attack.

6:05 PM

2. People from all over the country make calls on 115-Centre. The dispatcher shows the location of the blast incident. Everywhere around the vicinity is the fire ignition. The dispatcher initiates responses of the capital police officer, the city head of the Fire Brigade Department, the Emergency Medical Services Head, the city municipal head and the federal disaster manager. Calls are also sent for a helicopter request with its stabilised platform video camera.

6:25

3. Within minutes, the law enforcement personnel are on the scene. The SHO, Sadar with his team assumes the charge as incident command (IC). He identifies the scene and initially concludes that the blast is a suicide attack on the foreigners present in the hotel. He requests the dispatcher for a bomb disposal squad. Using GSP technology on electronic handheld devices, the IC receives a compiled roster of the bomb squad. The IC manages the scene through the inner perimeter and outer perimeter. Entry into the scene is allowed only to the authorized first responders. The officers start clearing the area.
4. Ambulances and fire brigades start arriving within minutes. The team of emergency medical services starts first aid under one commander.
5. Entry to the inner perimeter is stopped until it is declared safe.

7:45

6. Ambulances of the NGO and police start picking up bodies of the dead and wounded. Bomb disposal squads start arriving and conducting a chemical, biological, radiological and explosive sweep of the area.
7. Video cameras are mounted on the tops of vehicles of the police officers and fire brigades. These videos are transmitted live to the dispatch centre, fire battalion chief and other administration offices.
8. A unified command structure is established under the fire battalion chief; all are electronically connected through devices and the Internet.
9. The emergency medical services head initiates an emergency declaration in all city hospitals. On a GPS map display, he identifies the location for storage of emergency medicines on the scene. An inventory record of hospital bed positions is called on GPS devices by the medical head. The unified command calls the national database for qualified first responder volunteers. On arrival of these personnel, their authentication is through electronic passwords.

8:00

10. The bomb disposal squad declares the non-existence of any secondary device using chemical sensors or canine units. The IC permits first responders to enter the inner perimeter.

11. Firefighters mount cameras on their helmets and establish a mobile ad hoc network. Other high-bandwidth devices of short range are also connected and their communication with one another and with the command is established. Live images of the scenes are continuously transmitted to the command from helmet-mounted cameras.
12. From the hotel's surveillance camera, investigators download the immediate stored video and reconstruct events occurring a few minutes before the explosion. The camera shows a truck was approaching the hotel premises with the number plate AKZ 2893 and exploded. Investigators log into the national stolen vehicle database and find that the truck was stolen a day before the incident with the name of the primary suspect also identified. A photograph of the bomber was also downloaded and dispatched to NADRA authorities for identification.
13. Ambulances arrived and moved the dead and injured to hospitals.

6:00 AM (Next Morning)

14. Firefighters continued their efforts to control the fire. Fire extinguishers equipped with the necessary anti-fire and blast, anti-TNT and RDX materials were used to douse the fire till 6:00 AM of the next morning.
15. The last of the injured were transported to hospital.

7.4.3 Earthquakes

Earthquake early warning centres if established in the country, can predict an earthquake disaster minutes before it strikes the population. Research in earthquake engineering offers guidance that properly established earthquake early warning centres, through continuous monitoring of seismic station activities along the established historic fault lines, may give warning times on national channels of telecommunication. Not only this, but also the seismic activity sometimes generates ELF/VLF signals similar to one that caused severe earthquakes in the past, may be communicated to authorized centres for decision making of issuing early warning signals or not. This should be a monotonous process of research and, if continued for some period, may lead to "CONFIRMED" warning messages of even hours before an earthquake hits. However, we produce the scenario based upon the existing facilities the nation possesses and extract the communication needs of the future from that scenario. On October 8, 2005, at 8:50 local time, a 7.6 magnitude earthquake hit Muzaffarabad in Pakistan. Similar earthquakes with 7.6/7.7 magnitudes rocked San Francisco in 1906, Quetta in 1935 and Gujarat in 2001. The Government of Pakistan released information saying that the official death toll was 73,267, while officials say that around 1400 people died in India and that 4 people died in Afghanistan. The total death toll was over 74,500. The earthquake caused severe damage to lives, buildings and property, rendering homeless about 3.3 million people. The exact epicenter was about 11.8 miles northeast of Muzaffarabad and 65 miles northeast of Islamabad, the capital city of Pakistan. The hypocenter of the earthquake was located at a depth of 16.2 miles below the surface. The damage caused by the earthquake was devastating, the worst being in northern Pakistan, but also affecting other areas including Afghanistan, India and the southern parts of the Kashmir Valley. Islamabad and Karachi felt an aftershock of 4.6 on the Richter scale. There were 147 aftershocks on the first day after the initial shock. One of the highest aftershocks was recorded at 6.2. There were 28 aftershocks that had a higher magnitude. On October 19, 2005, a series of aftershocks were also felt, one with a magnitude of 5.8; this was located 40.5 miles above Muzaffarabad. There were more than 978 aftershocks recorded, all of which occurred daily and continued until October 27, 2005.

8:40 AM
Gas lines in and around the city as well as railway engine drivers receive signals for the automatic shutoff of supplies through automatic sensor systems. All emergency call channels either suffered damage or were overloaded due to public enquiries. Cellular systems were minimally operational. The scenario shows the communication needs of the country during the response phase of such a major disaster.

1. The Civionics Earthquake Early Warning Centre (CIVEEWS) situated in Rawalpindi received a P-wave signal from various seismic monitoring stations and generated an early warning alert message of severe earthquake of magnitude 7.6 to the entire country's warning centres, services chiefs, police department heads, public (of the affected area) mobile phones (with evacuation messages), firefighting cells, authorized first responders, hospitals, police stations, legislators, NDMA chief, all heads of resources, NGOs, volunteers and to all those whose phone numbers are there in their database and are connected to the early warning message dispatcher. Alert confirmation is made via telephone, voice radio and on TV channels.
2. Connected people who received messages left their premises; however, many could not due to winter sleep and remained under the debris.
3. Disaster responders across the country were called as the disaster hit the city through wired/wire-line/messages and TV/radio appeals for volunteers.
4. Reports of deaths and injuries were received from the citizens. Reports of buildings that collapsed, that were near to collapse and that were safe were received. Everywhere, there were scenarios of screams and cries.
5. City-to-city communication is restored via the satellite system.
6. The president, prime minister, ministers, services chiefs, secretaries, heads of departments, and so on, were connected into conference calls through the VSAT system. Unified response and rescue strategy was discussed and resolved.
7. GIS-GPS-enabled PDAs were dispatching disaster site videos to all responsible commanders and heads.
8. The GIS-GPS-based database was backed up from other provincial head offices to localize disaster sites and estimate losses. The database also indicated the locations of various infrastructure, hospitals, offices, firefighting resources, roads, buildings, residential and commercial areas, etc.
9. Satellite photographs and aerial images were transmitted to local heads.
10. A trailer-mounted high-capacity satellite terminal was deployed in the city.
11. Public safety wireless devices were distributed among the core first responders including the Red Cross team responsible for arrangements of shelter.

These devices have the capability to survey the buildings' structural integrity, victim locations, set up wireless MANET and were linked with the central GIS database to share voice, video and data with other similar devices. With these devices, the medical teams tracked the earthquake victims. These devices also possessed a prioritization option by virtue of which the voice, video and GIS data were given appropriate priority to avoid congestion of or demand for any particular type of data.

12. The structural specialists arrived and established theodolites to monitor any movement in buildings of critical importance. These were accompanied by video cameras to transmit information of building images through wireless medium to the team of experts.
13. The demand for estimation of exact losses increased. Computer-based loss estimation tools (like Hazus-MH) were used to calculate losses.
14. The Sahana Disaster Management Tool was immediately configured and roles were assigned in PWDs to all first responders, volunteers, donors, hospitals, patients, national and international NGOs.
15. Displaced victims were transferred to the disaster city for internally displaced (DiCFID).

7.5 REQUIREMENTS FOR AN INTEROPERABLE COMMUNICATION NETWORK

An interoperable communication network for disaster management must possess parameters of availability, security, quality of service (QoS) and the capability to operate in a mobile environment. A brief introduction to these parameters follows.

7.5.1 AVAILABILITY A_0

Availability refers to the theme that the network is present without failure in a 24/7 time duration. Continuous availability is the main criterion for an efficient communication network. For a reliable system and infrastructure, the maximum possible availability is required. The three requirements of high availability engineering are: no single point of failure, a consistent crossover and the ability to find the point of failures. "Availability" also refers to the "ability of the system in which it is being fixed up in time units and is up and re-functional."

In contrast, reliability refers to whether or not the infrastructure is working; capability is the sum of services provided for the first responders and scalability refers to how well the system handles surge conditions, whereas survivability measures the resistance of the system to failure, and finally, there is restorability, which is a measure of how easy the system is restored from failure [3].

If networks pose the problem of failure, their availability is weak and such networks are not acceptable. Using commercial-off-the-shelf (COTS) technology, networks can be designed with maximum availability. Availability is defined as under [3]:

$$A_0 = \frac{\text{up time}}{\text{total time}} \tag{7.1}$$

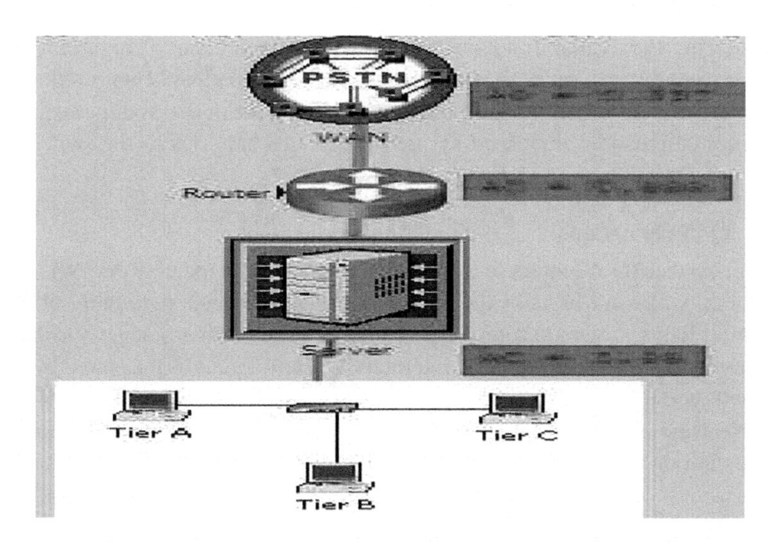

FIGURE 7.2 Availability calculation example.

In a network, the attached components have varying availability linked together. In Figure 7.2, different components of the network have varying availability values. In this figure, the availability of WAN, router and LAN is 99.7%, 99.9% and 99%, respectively. The overall availability of the network is calculated as

$$A_0 = 0.997 \times 0.999 \times 0.99 = 0.986$$

In a month with 30 days, there are 43,200 minutes. The network in Figure 7.2 will be up for 42,595 minutes and down for 605 minutes, which is a relatively long downtime period. To resolve this issue, it is best to adopt redundancy. Now consider the network in Figure 7.3.

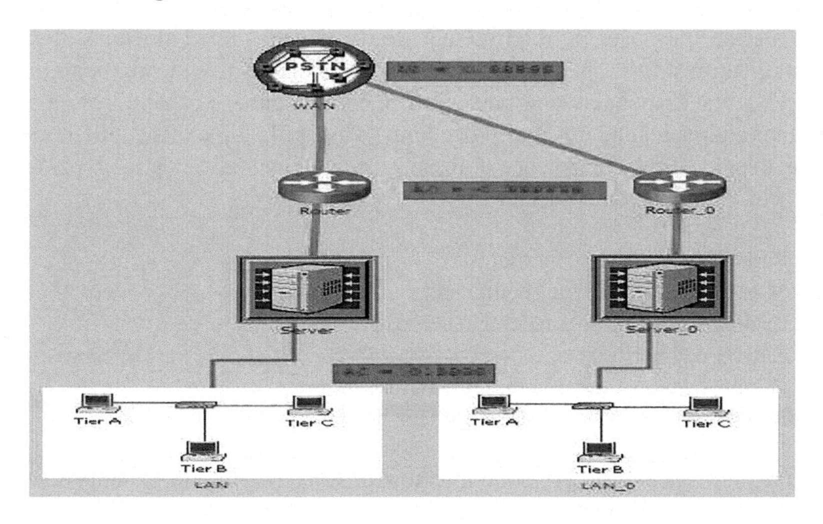

FIGURE 7.3 Example of availability calculation with different component values.

Considering the availabilities of the individual components of Figure 7.3 as we did in the previous example, the overall availability is 0.999889 with a downtime equal to 5 minutes, which is a remarkable improvement in the system availability. Availability can also be increased by providing extra batteries to the switches and routers.

7.5.1.1 QoS versus qos

Quality of service (QoS) refers to controlling congestion in the network. WiMax can control QoS, while Wi-Fi lacks the ability to control congestion. Quality of service (qos—small letters) refers to some factors such as bandwidth efficiency (throughput) [1,3] determinism, latency, jitter and interactivity. Throughput is measured by packet per second, while latency is the time it takes for a packet to get through a network. Jitter is the time of delivery of packets from a source to a destination. These factors are dependent on each other.

7.5.1.2 Security

Security is categorized into two parts [3]: content protection and infrastructure protection. Content protection includes authenticity, confidentiality, integrity and no repudiation quartet. Infrastructure protection includes resistance to denial of service (DOS) and type of service (TOS) attacks, resistance to threat analysis (TA)/two-factor authentication (computer security authentication) analysis, low probability intercept (LPI) and transmission security (TRANSEC) issues.

Other attributes for a network are coverage, accessibility, transmission speed, fast call setup, network capacity, direct local communication, addressing functionality, cost and mobility.

7.6 PAKISTAN'S DISASTER COMMUNICATIONS INTEROPERABILITY PLAN: A PROPOSAL

This section proposes a requirement analysis plan for Pakistan's disaster communications interoperability. The motive behind this work is to make Pakistan's first responders and law enforcement personnel's capabilities on a par with international standards and procedures. This will also enhance the response preparedness and the competence of these agencies. Generally, any interoperability plan includes in it features such as [3]:

- Redundant transmission routes
- More sophisticated vital equipment
- Backup subscriber management centre
- Emergency hotlines
- Paging systems for spreading disaster warnings
- Rapid upgradation techniques for base stations during response

The core of this plan is the IP-based infrastructure and the remaining setup is linked with the core via gateways. The selection of equipment will be made keeping in view the available budget, technology and the country's political scenarios. The plan

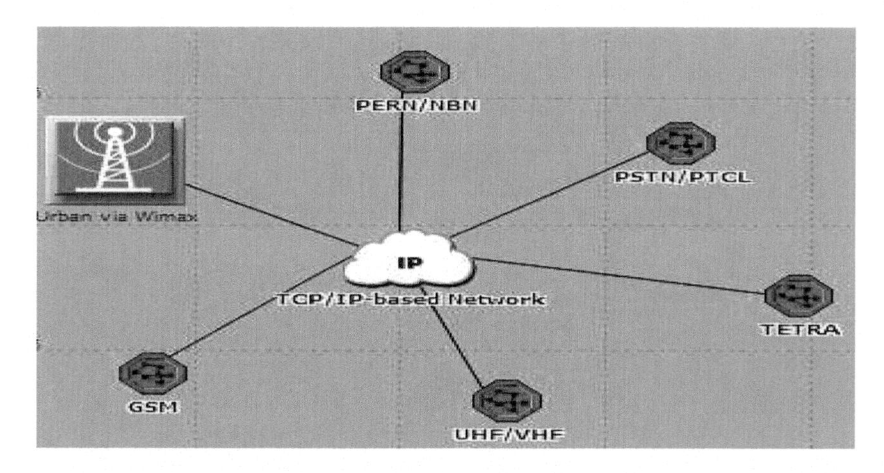

FIGURE 7.4 Proposed TCP/IP-connected network.

is based upon the theme that the existing circuit switching network, along with the Internet infrastructure, does not become obsolete; instead, it should be linked with the new homogenous packet-switching setup using gateways as the interfaces. Through the proposed plan, the communication between the citizens and government and between intra-government components is made easy and is also scalable.

In Pakistan, the law enforcement agencies rely on their own UHF/VHF systems; it is recommended that they should immediately switch to the TETRA System which is implemented worldwide for public safety operations [1]. It has been proposed that the TCP/IP-based backbone should be linked to other networks like National Broadband Network (NBN)/Pakistan Educational Research Network (PERN), Public Switched Telecommunication Network (PSTN), which is fiber based, GSM, UHF/VHF using gateways (Figure 7.4). The remote and urban areas can be linked via WiMAX and Wi-Fi fringes.

It has also been proposed that by using satellite phones (manufactured by Thuraya, BGAN or Iridium), the country's first responders can be linked with

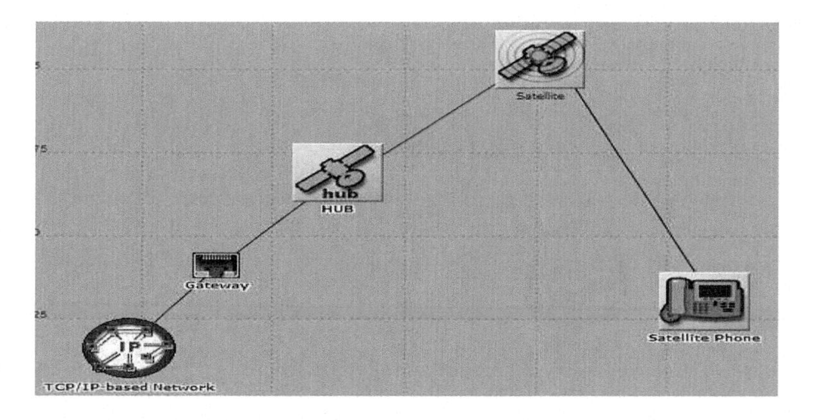

FIGURE 7.5 Infrastructure for TCP/IP satellite phone connectivity.

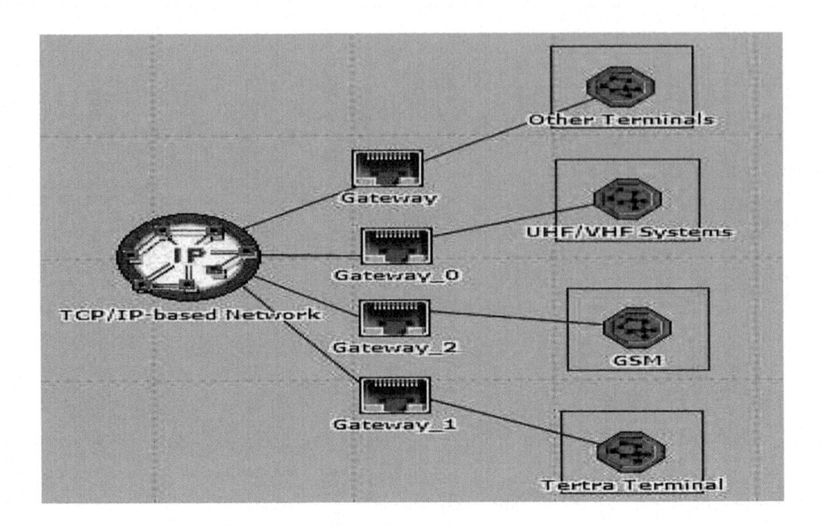

FIGURE 7.6 Infrastructure for TCP/IP connected through gateway with other networks.

foreign communication systems or with networks across the city through the same TCP/IP-based networks (Figure 7.5). The same concept is illustrated in Figure 7.6, wherein the subnet "other terminals" refers to either PERN or schools or hospitals or other offices, as well as both state-owned and private NGOs that can be linked to the system through appropriate IP-based gateways before, during and after the disasters. In designing all these systems and interoperable networks, redundancy in links is the core theme and wherever the failure of one link is determined, switching to the other link must be restored. Generally, the WiMAX systems with Wi-Fi fringes have the best capability to establish redundancy in coordination with satellite connectivity. This is

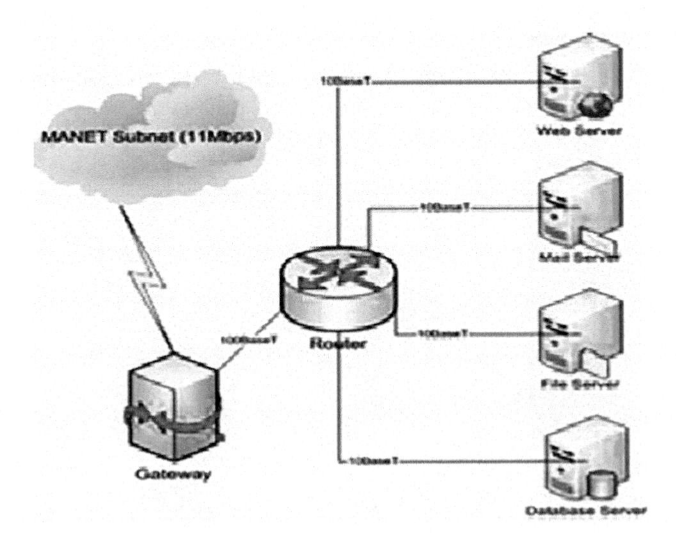

FIGURE 7.7 Interconnection of fixed network with a MANET subnet through a gateway [6].

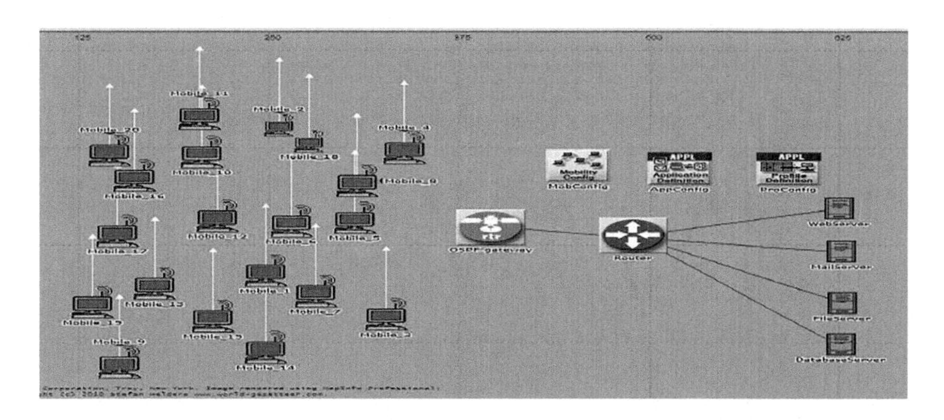

FIGURE 7.8 Interconnection of a fixed network with a MANET through a gateway—OPNET scenario [6].

also needed during the response phase of disasters when most of the fixed infrastructure is damaged.

7.7 SIMULATION OF SCENARIOS

To validate the above-mentioned discussion, various scenarios have been considered and simulated in the OPNET Modeler 14.5 [1] environment using the AGI Satellite Toolkit (STK) [4].

7.7.1 Scenario I: Fixed Network Interconnected with a Mobile Ad Hoc Network

In the first scenario, we have shown interconnecting mobile ad hoc networks with legacy networks using Internet gateways [5]. The simulation shows results for a fixed radius of nodes with varying pause times at different mobility levels for measured QoS parameters of packet loss, routing traffic received/sent and total replies received from destinations using the random waypoint propagation model. The results prove medium level satisfaction over decreased packet losses, increased packet transmission and their acknowledgement with both lower and higher mobility levels for varying pause times. The core issue is to identify the proper Internet gateway and associated protocols to support the necessary services. In these types of networks, the fixed infrastructure lies at one end and the mobile at the other [6].

DISCUSSION ON RESULTS

During the simulation phase of this research, 20 nodes are selected for each scenario with a radius of 100 metres and the propagation model is the "Random Waypoint" in the OPNET Modeler environment [1]. With a constant radius in each scenario, maximum speed and pause time are varied. For each scenario, speeds of 2 (pedestrian), 15, 30, 45 m/s are used with pause times of 0, 15, 30, 45 and

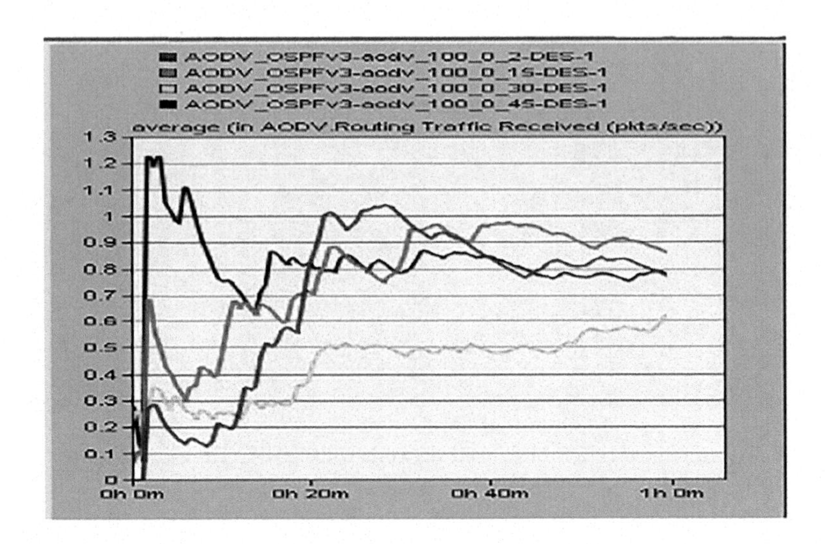

FIGURE 7.9 Routing traffic received—pause time = 0 second [6].

100 seconds. The AODV routing protocol is run over OSPFv3 using the Internet gateway router. OPNET provides several options to compare and analyze a variety of scenarios. However, important attributes over which the performance is studied in this work are total packets dropped, routing traffic received (packets/s), routing traffic sent (packets/s) and total replies sent from destination. The simulations are run for a period of approximately 3600 seconds and results are compiled as shown in Figures 7.9 through 7.24. The results conclude how AODV performs/behaves over OSPFv3 in the different scenarios [5].

From Figures 7.9 through 7.12, it is evident that for higher speeds all the selected parameters such as routing traffic, total replies sent from destination and packet loss

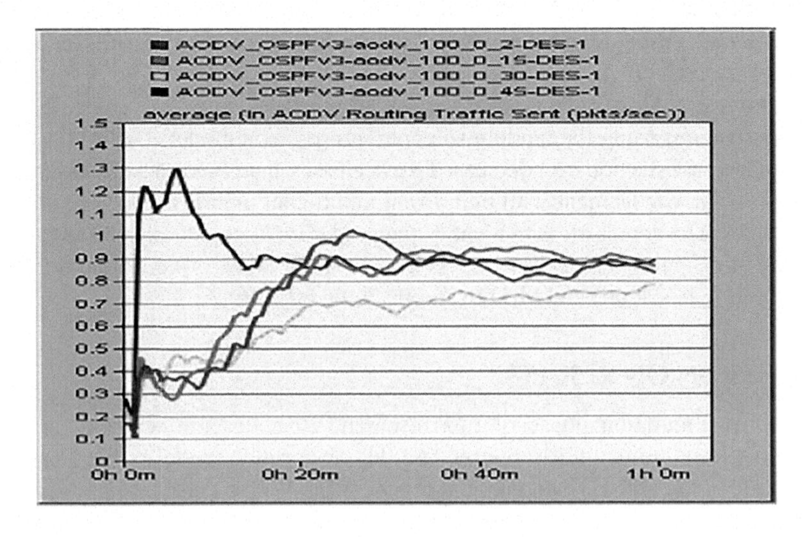

FIGURE 7.10 Routing traffic sent pause time = 0 second [6].

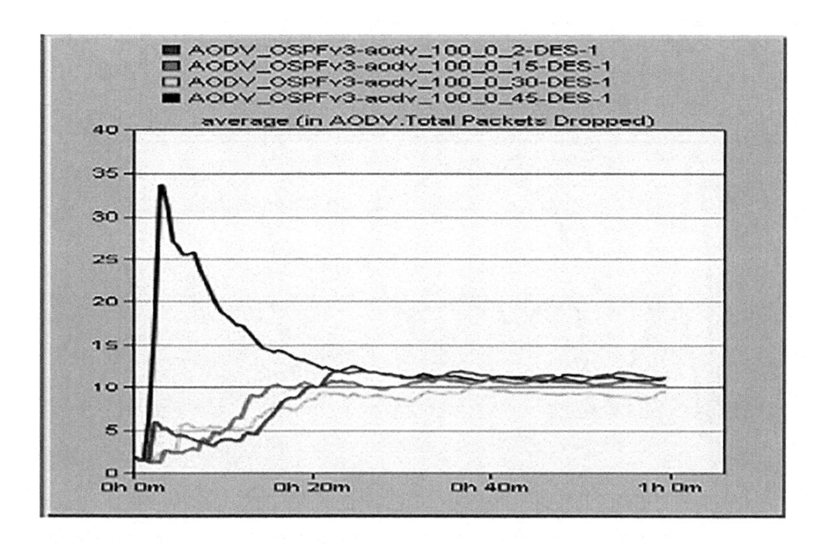

FIGURE 7.11 AODV total packets dropped [6].

are also showing a rapid rise for some initial periods, but after some time relative stability is observed. This is valid when the "radius" and "pause time" are constant, that is zero, but the "speeds" are varied.

The next four results from Figures 7.13 through 7.16 are graphs when the pause time is fixed as 50 seconds with the same radius and varying speeds. In these four cases, an initial rapid rise in routing traffic, packet drop and total replies sent from destination is noted for the least speed that is 2 m/s. In the scenario in Figure 7.13, it is also interesting to see that that for all selected speeds at the pause time of 50 seconds, the response of traffic received per second is good enough; whereas for the traffic sent, Figure 7.14 is medium. However, the replies sent from destinations are slow at

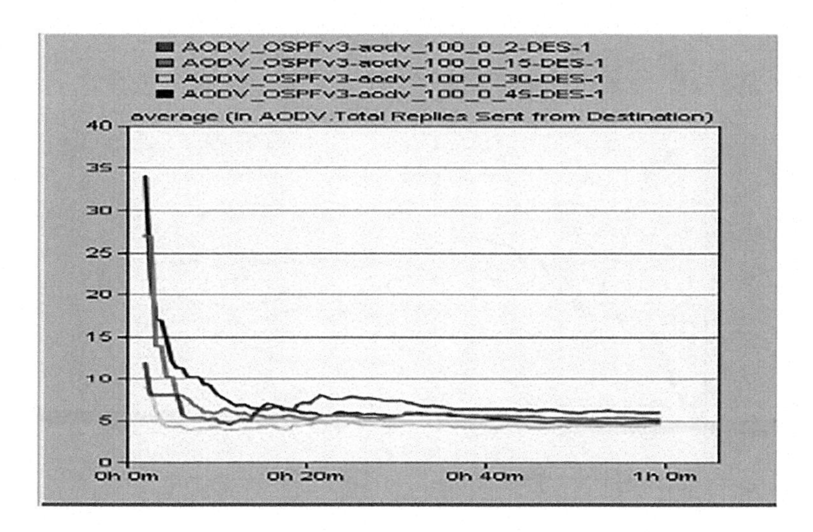

FIGURE 7.12 Total replies sent from destination pause time = 0 second [6].

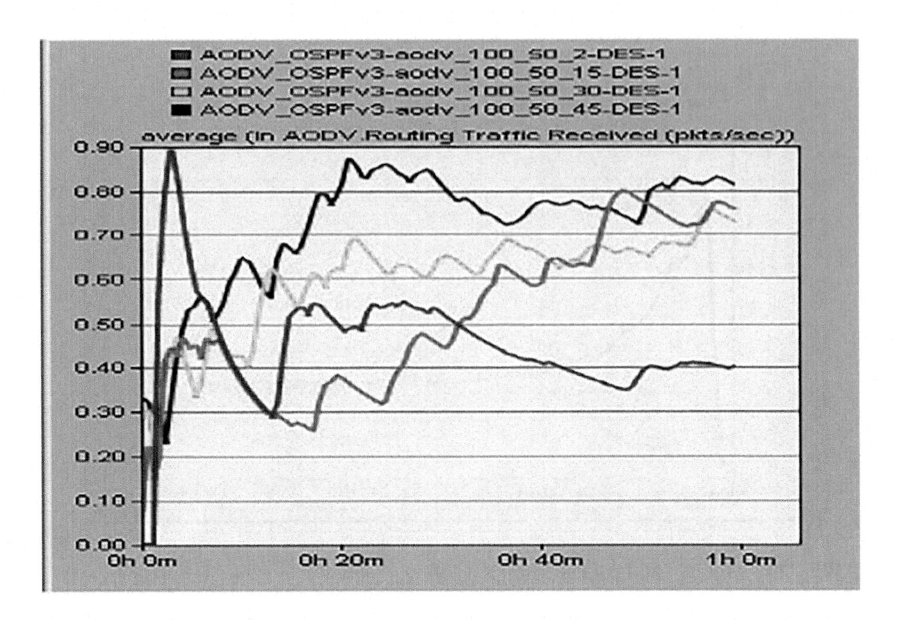

FIGURE 7.13 Routing traffic received—pause time = 50 seconds [6].

speeds of 45 m/s due to the heavy traffic load and protocol overhead of AODV over OSPFv3 using the Internet gateway router. When the pause time is set as 100 seconds, the results in Figure 7.17 indicate a rapid initial jump for received routing traffic at a speed of 30 m/s and fast sent routing traffic at the least speed of 2 m/s (Figure 7.18). This result is also encouraging as far as protocol performance of AODV and OSPFv3

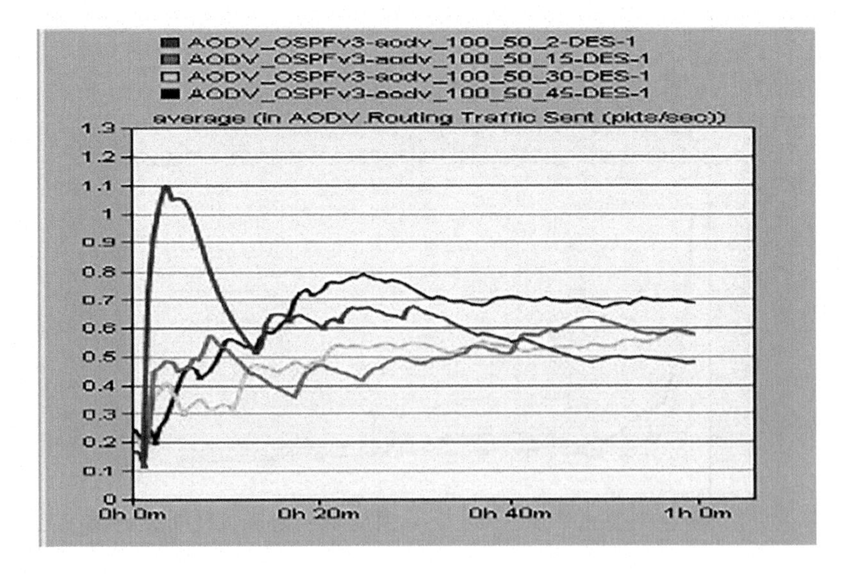

FIGURE 7.14 Routing traffic sent pause time = 50 seconds [6].

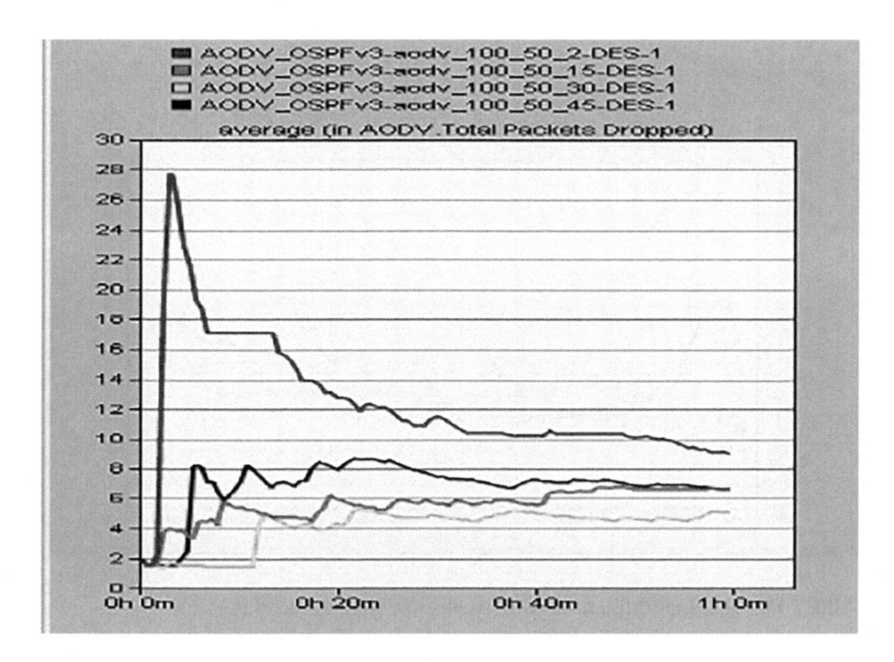

FIGURE 7.15 Total packets dropped—pause time = 50 seconds [6].

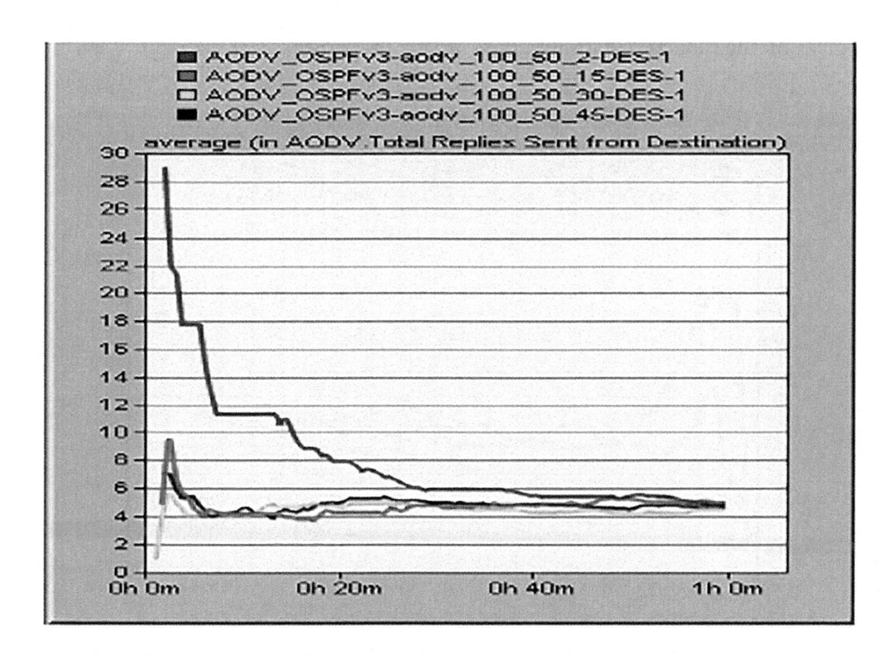

FIGURE 7.16 Replies sent from destination pause time = 50 seconds [6].

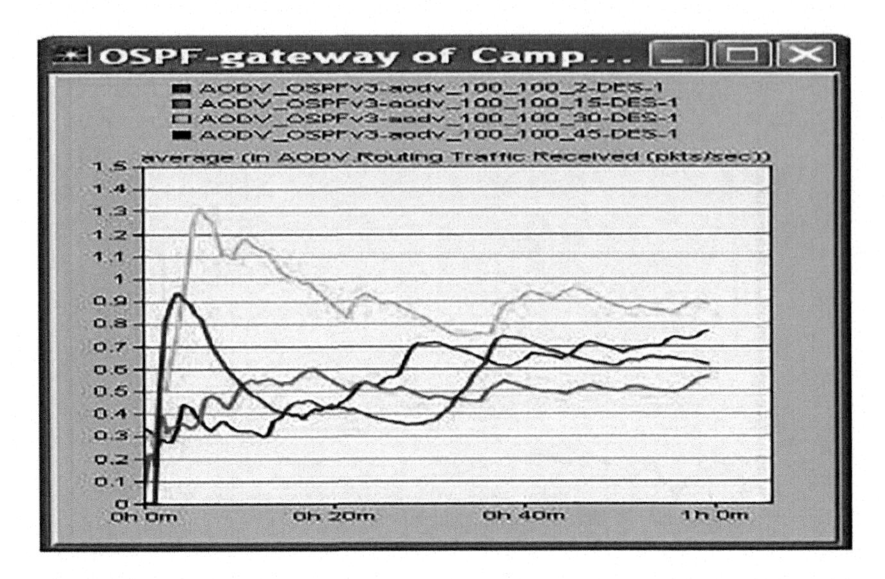

FIGURE 7.17 Routing traffic received—pause time = 100 second [6].

is concerned in a dynamic environment. Also for higher speeds (Figure 7.19), that is 45 m/s, the ratio of packets dropped is the least as compared to a lower speed, which is also a healthy sign but due to congestion in the network, responses received from the destination are poor for higher speeds and a little bit encouraging at lower speeds, particularly at 2 m/s (Figure 7.20).

For the highest pause time selected in this research, that is 250 seconds, the response of the selected parameters is given in Figures 7.21 through 7.24. The

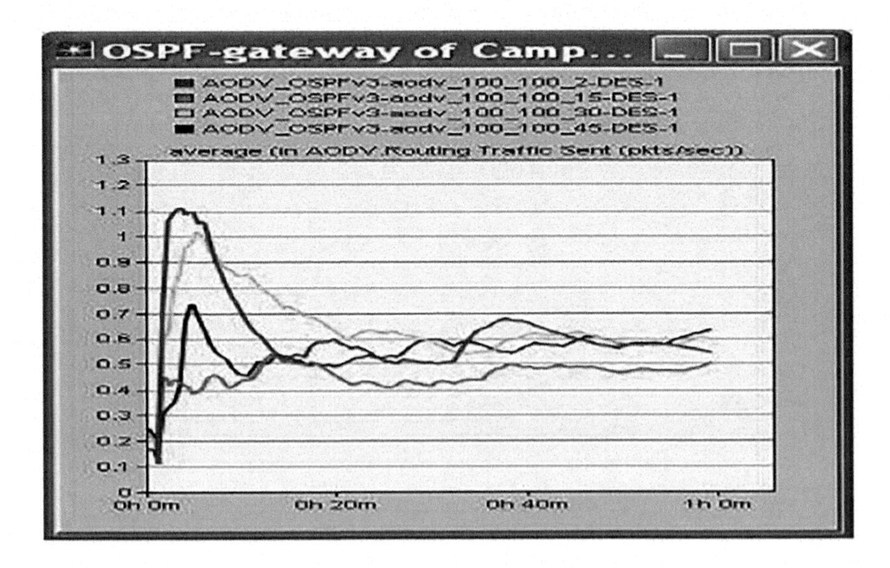

FIGURE 7.18 Routing traffic sent pause time = 100 seconds [6].

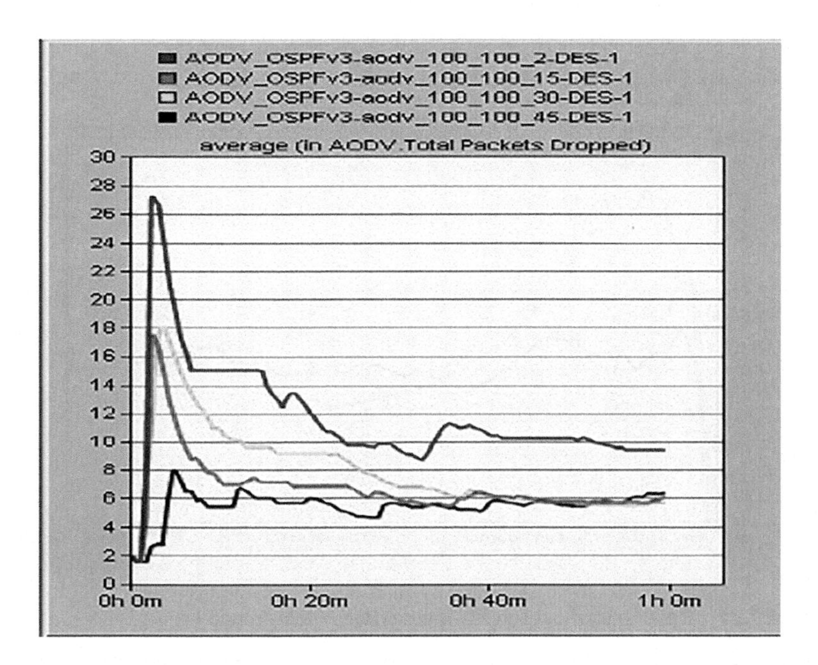

FIGURE 7.19 Total packets dropped—pause time = 100 second [6].

received AODV routing traffic for a speed of 30 m/s is fast enough to be observed in Figure 7.21. However, the ratio of packet loss is medium for higher speeds like 45 m/s (Figure 7.22). The response received from the destination at high speeds (30 m/s) is also rapid (Figure 7.24) but this appears to be decaying after the passage of time for all selected speeds due to traffic load and congestion in the network.

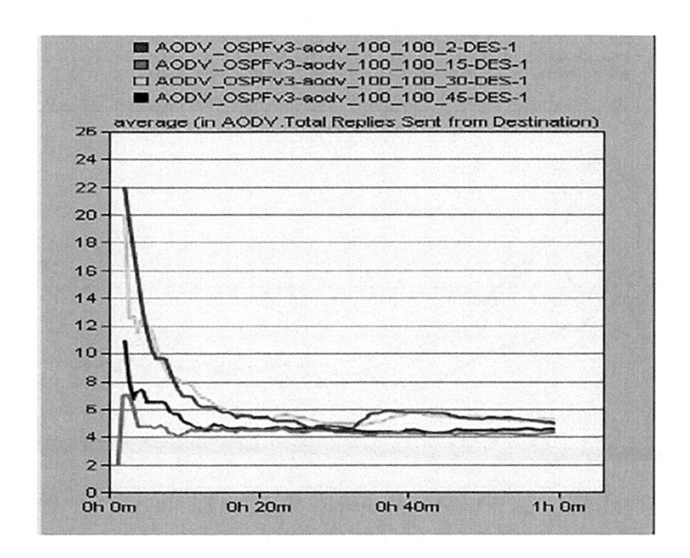

FIGURE 7.20 Replies sent from destination pause time = 100 seconds [6].

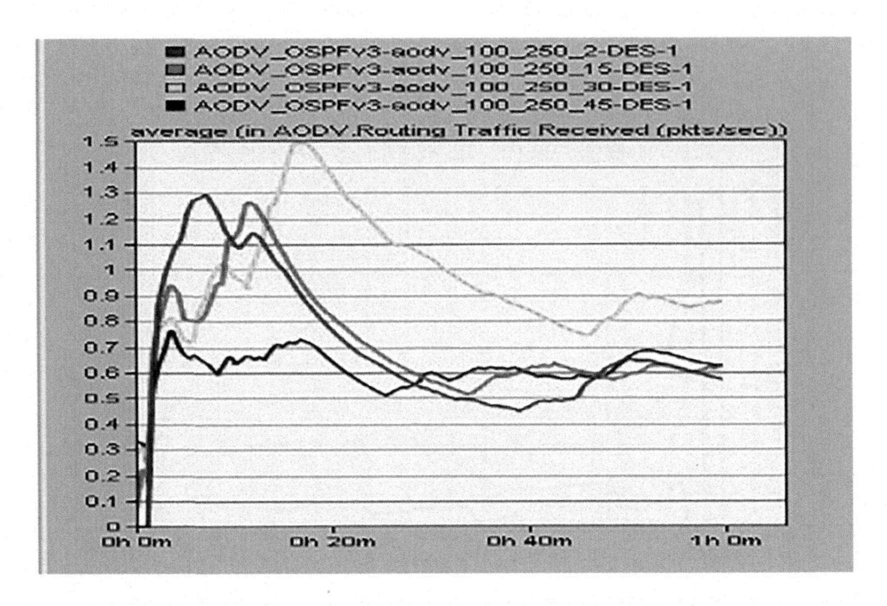

FIGURE 7.21 Routing traffic received—pause time = 250 second [6].

In the cross-domain environment of interconnecting mobile ad hoc networks with fixed infrastructure-based networks using routers in the form of Internet gateways, various routing protocols are currently being used at either side. One of the solutions proposed in this research is that the AODV routing protocol over OSPFv3 can be the reliable candidate for heterogeneous ubiquitous environment. The experiments in a simulated environment for such a solution were performed through lab tests using

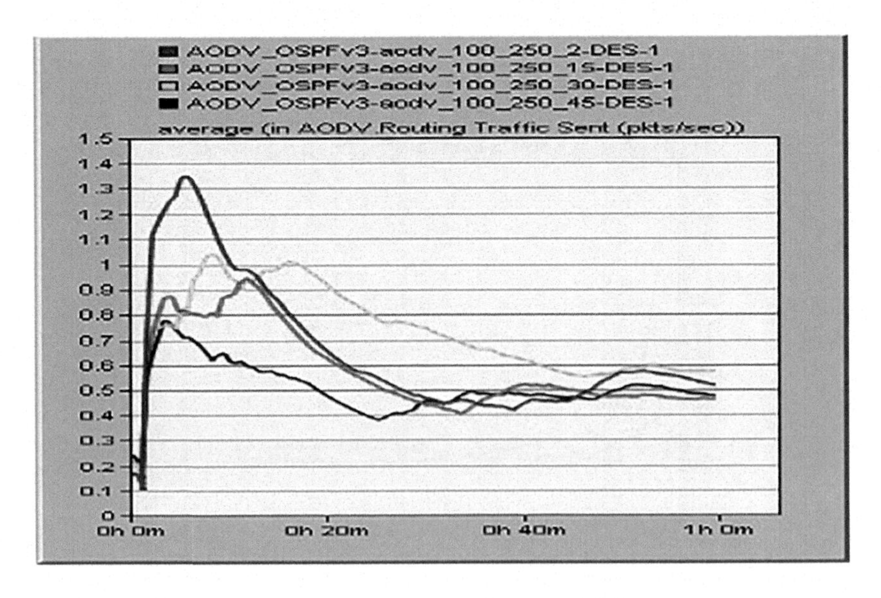

FIGURE 7.22 Routing traffic sent pause time = 250 seconds [6].

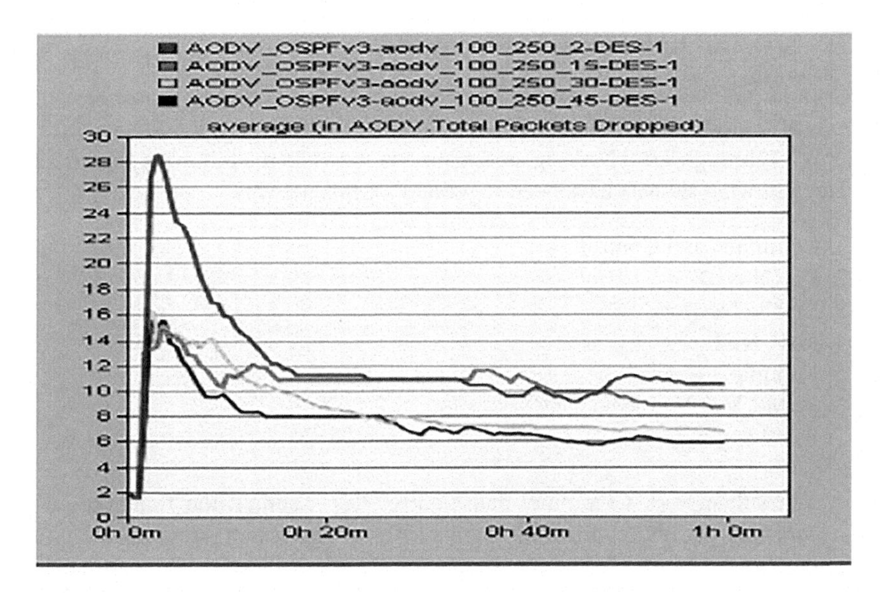

FIGURE 7.23 Total packets dropped—pause time = 250 seconds [6].

the OPNET Research and Development Modeler Software. The simulations show a satisfactory performance of AODV over OSPFv3 routing protocols for the network of 20 nodes with a radius of 100 metres and varying mobility levels and pause times using the random waypoint propagation model. The results also provide a satisfactory decreased ratio of packet loss during transmission and reception for low, medium and high traffic scenarios of varying pause times and speeds [5].

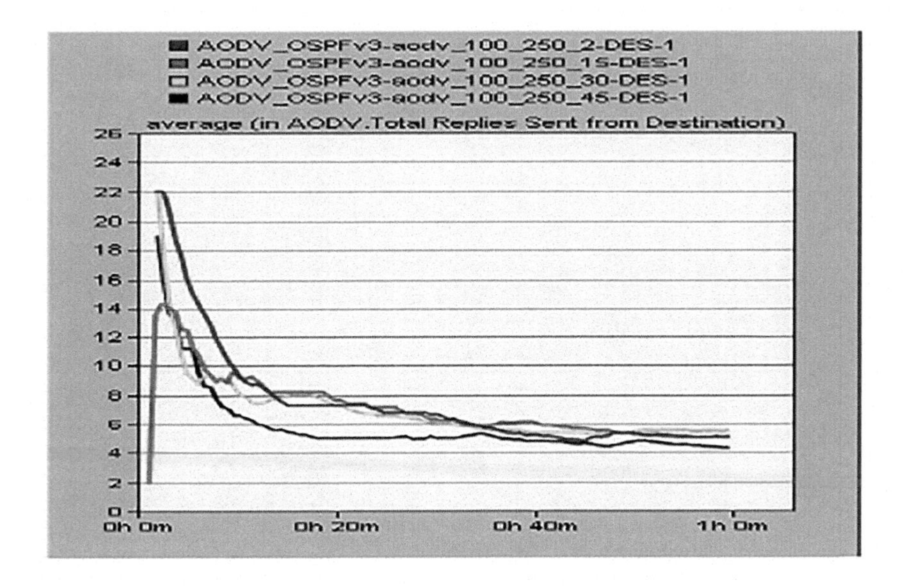

FIGURE 7.24 Replies sent from destination pause time = 250 seconds [6].

7.7.2 Scenario II: Interoperable Networks

In a similar manner as discussed in Section 7.7.1, this section presents the design of the scenario in an OPNET-simulated environment matching the criteria as set forth in Figures 7.4 through 7.6. The designed scenario is shown in the following Figure 7.25.

The following subnets have been suggested for this network:

1. Command and Control
2. PERN/NBN
3. Mobile Network
4. MIP_NET
5. Multiple Agencies
6. WiMAX Network
7. Satellite Operations

1. The subnetwork Command and Control has assumed the functions of coordinating and controlling the agencies of the country responsible for the disaster management department. The National Disaster Management Authority, SUPARCO, can be the headquarters of this subnetwork, which is electronically linked with the country's other networks on the IP-based communication system. The generic architecture of this subnetwork is shown in Figure 7.26.
2. PERN/NBN stands for Pakistan's Educational and Research Network/ National Broad-Based Network. This subnet has assumed the role of

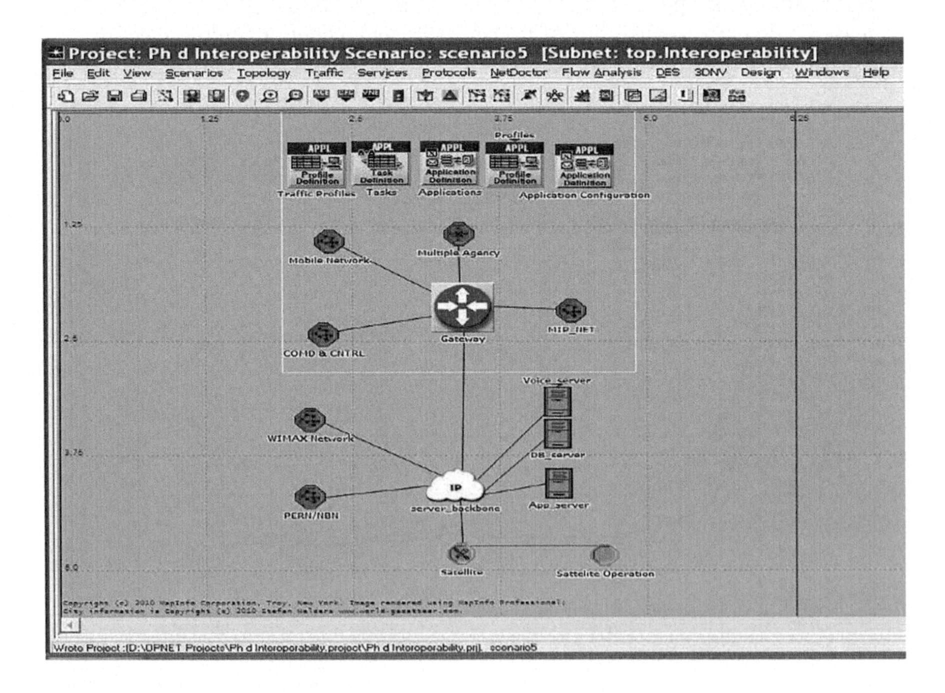

FIGURE 7.25 Proposed interoperability model designed in OPNET.

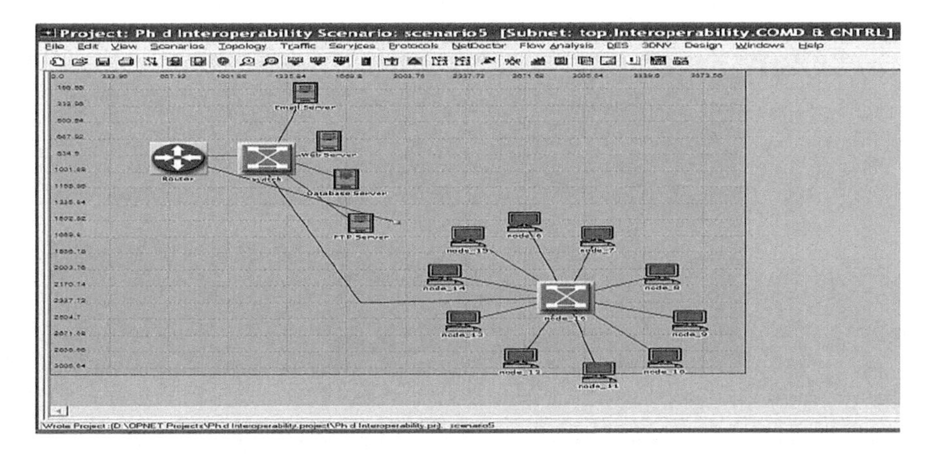

FIGURE 7.26 Component of interoperability scenario—command and control.

connecting the country's existing core public infrastructure such as universities, hospitals, educational institutes and NGO offices through fiber links. Naturally, these departments play a considerable role before, after and during the disasters. The scenario subnet is shown in Figure 7.27.

3. A mobile network is a scenario based on VoIP technology. This type of connectivity is also necessary for tackling the disaster management issue where we lack fiber or satellite connectivity. In this scenario, the network

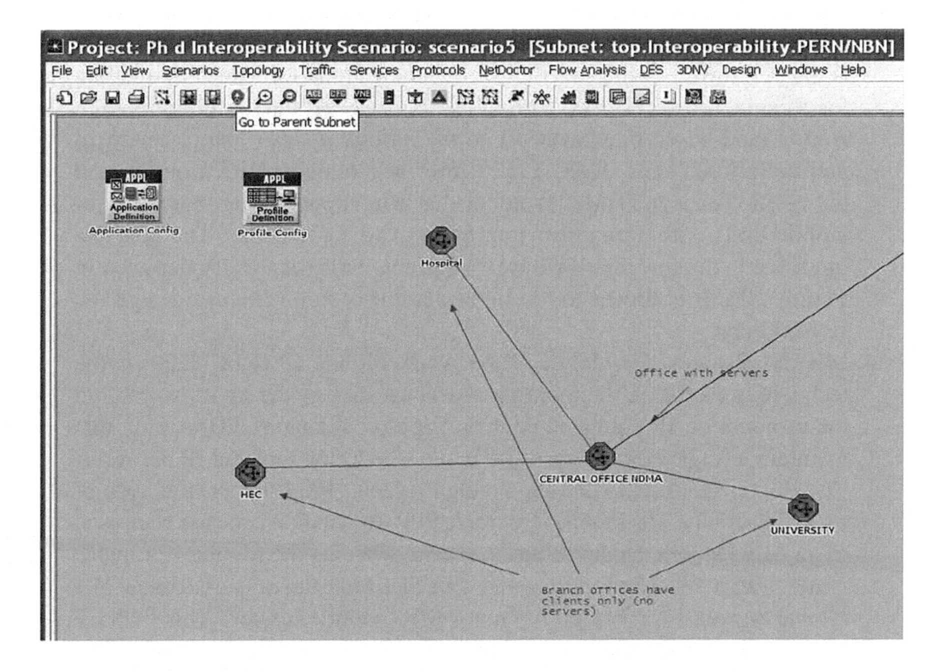

FIGURE 7.27 Component of interoperability scenario—PERN/NBN.

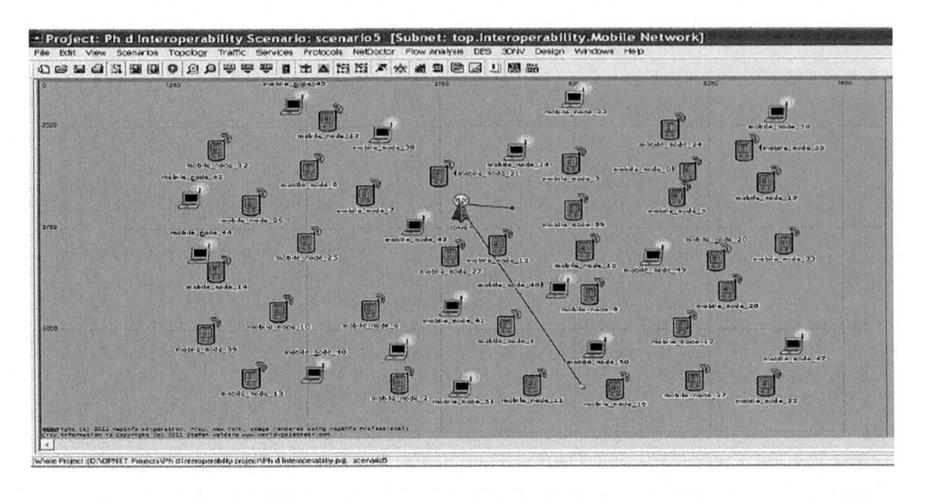

FIGURE 7.28 Component of interoperability scenario—mobile network of first responders.

performance of a cell supporting 49 mobile clients has been shown. Each type of node has a different mix of network applications including FTP, web browsing, email and database sessions. These nodes represent a mobile laptop or a handheld data unit and a VoIP mobile phone. Each device has data access through the local cell tower. These nodes use the standard protocol stack (Application/TCP/IP) of a workstation in the OPNET standard model library. These client nodes will access the tower that is in the same subnet via a TDMA radio interface. This node is the cell tower. Using two radio frequencies (one for Transmit and one for Receive), the cell tower provides two-way IP data communication from a land-based fixed network to the mobile nodes (laptop, handheld and cellphone). The MAC Layer interface between the mobiles and the cell tower is a TDMA protocol that can support a large number of mobile units with data rates from 64 kbps to 1.472 Mbps. The landline interface to the tower is 100BaseT Ethernet. This scenario is depicted in Figure 7.28. It is linked to the Interoperability main router through the IP backbone.

4. MIP_NET stands for Mobile IP Network. Mobile IP (RFC 3220) is the technology based on the roaming characteristics of the nodes which run the protocol on IP equipped devices (laptop computers, PDAs, etc.) that maintain a single point of presence while wandering on other IP networks. The Home Agent (HA) and the Foreign Agents (FAs) support this type of roaming activity. The Mobile IP capable PPP workstation node has been used as an MN (Mobile Node) in this scenario, whereas the Mobile IP capable router with a WLAN, Ethernet and four SLIP interfaces is used as an HA (Home Agent), FA (Foreign Agent) or MR (Mobile Router). This OPNET scenario has been imported so as to show the presence of various agencies

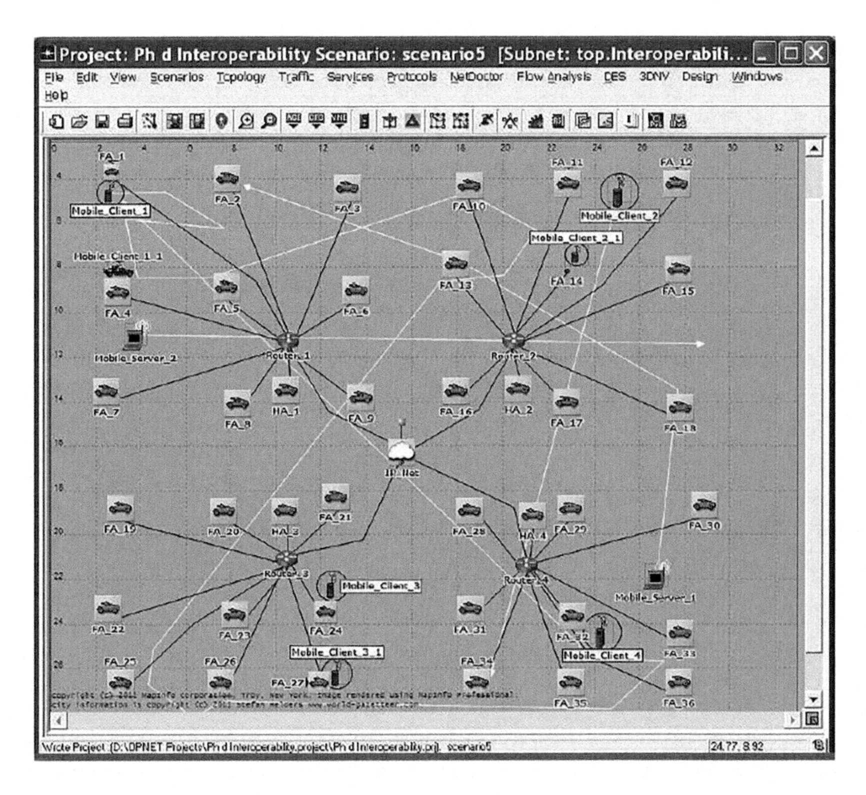

FIGURE 7.29 Component of interoperability scenario—mobile IP network of first responders during response.

on the spot of a large disaster wherein there appears to be a destroyed fixed network. The sense of communication existence in such a situation can be made using mobile devices, laptops and PDAs linked with each other and with the Internet using the MIP_NET concept given therein. This scenario is shown in Figure 7.29, which has been linked to the interoperability subnet through the IP router.

5. Multiple Agency Network: The concept as shown in point 4 above can be extended for a multiple agency network requirement that can be mobile or fixed with roaming as well as non-roaming attributes with connectivity of all these nodes through a central hub and wireless access points as shown in scenario Figure 7.30. The need for intercommunication with mobility characteristics having a central command and control on the disaster site is the paramount factor to handle the worse situations efficiently while maintaining communication with the central command and other networks.

This OPNET scenario is an 89-node WLAN network with varying WLAN algorithms and features specified in IEEE 802.11 and 802.11b standards, like various WLAN data rates, RTS/CTS frame exchange, data

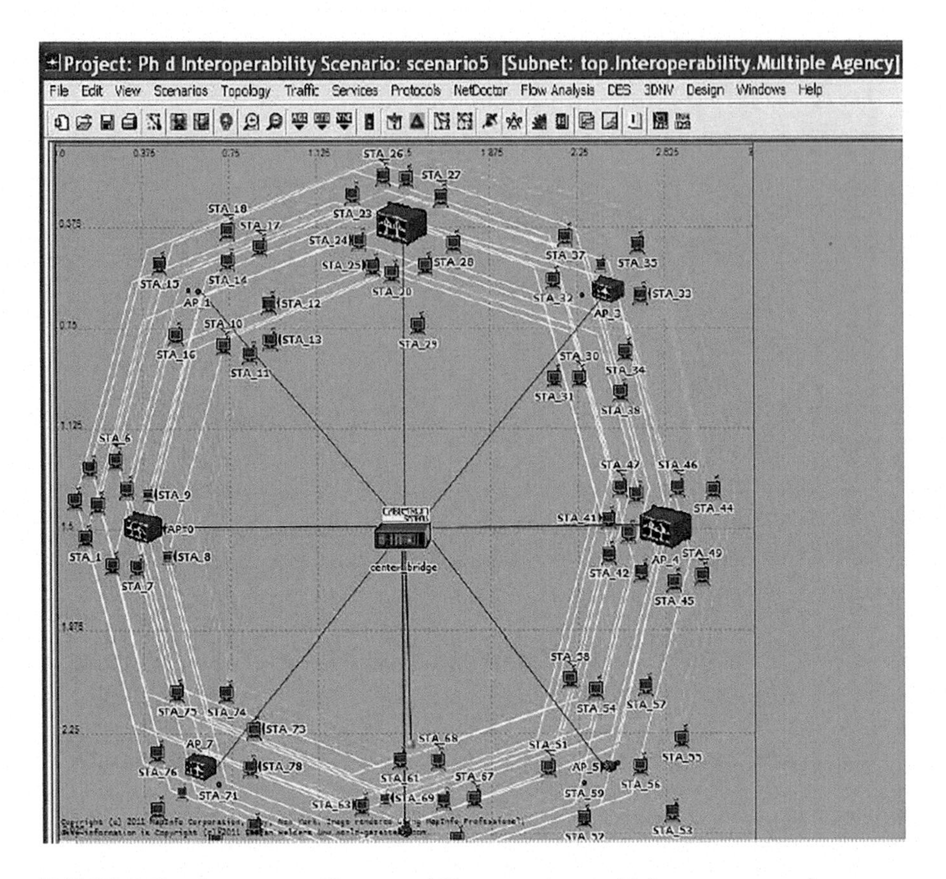

FIGURE 7.30 Component of interoperability scenario—multiple agency network.

packet fragmentation and roaming. In this scenario, some of the nodes can transverse their trajectories but still can maintain communication with their parent access point. This type of situation is very much needed in the disaster response phase when multiple agencies with varying devices need to adopt roaming in other types of networks while maintaining connectivity not only with their own nets but also with other systems.

6. WiMAX Networks: The concept of WiMAX technology has already been discussed in designing an early warning system in earlier chapters wherein it was stated that at the urban sites or during the scenarios when total collapse of infrastructure has taken place, the WiMAX solution is preferable to establish prompt communications. The WiMAX scenario is repeated here for interoperability purposes and is illustrated in Figure 7.31.

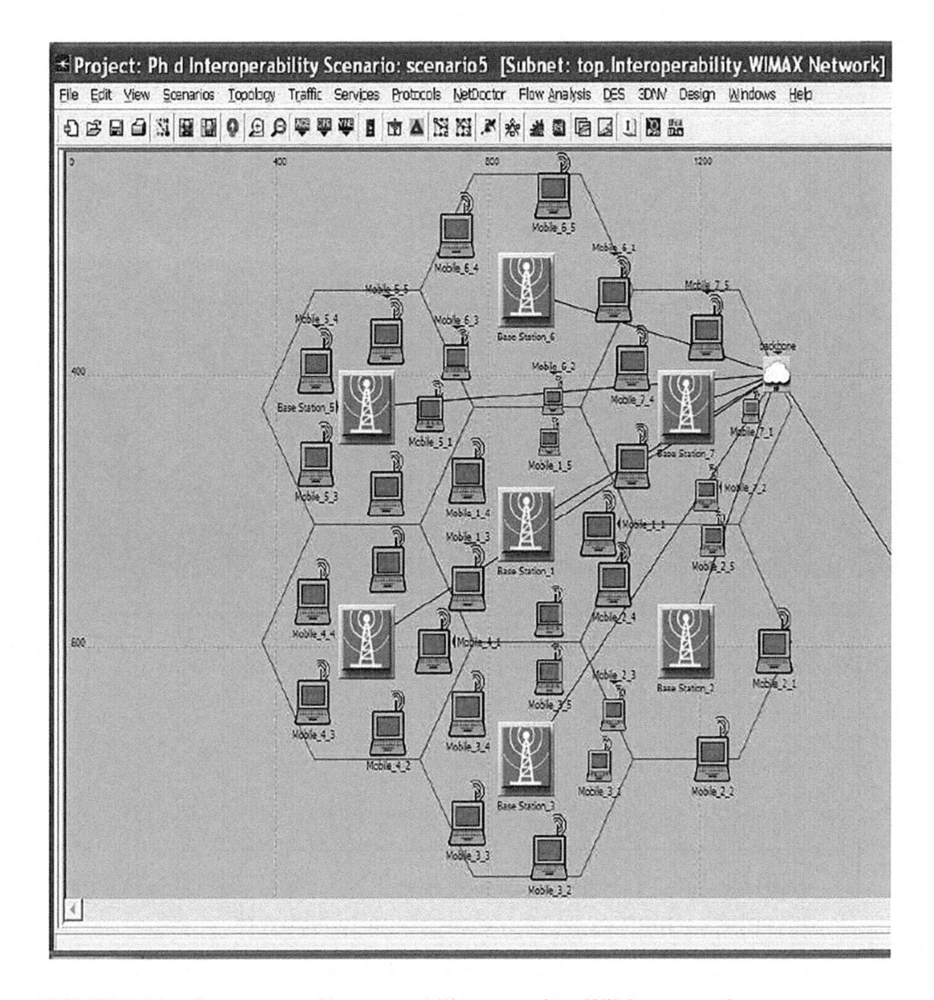

FIGURE 7.31 Component of interoperability scenario—WiMax network.

7. Satellite Operations: Similar is the case of satellite connectivity as shown in the following scenario when one remote operation is being made with the IKONOS orbit satellite as shown in Figure 7.32. The above-mentioned scenarios 6 and 7 are linked to the main interoperability subnet through the IP backbone router and Internet (Table 7.2).

7.7.3 Results of the Interoperability Scenario

Figure 7.33a depicts the 1 Simulation progress of the interoperability scenario and its results are shown in Figures 7.33b through 7.48. These relate to the proposed system's operational activity, link quality, packet status, network survivability and overall performance of the network, including congestion in the channels.

FIGURE 7.32 Components of the interoperability scenario—satellite operations.

TABLE 7.2
Inventory Summary of an Interoperability Network

S. No.	Element	Type	Count
1	Devices	Total	289
2		Routers	70
3		Switches	5
4		Layer-3 Switches	1
5		Hubs	4
6		LAN Nodes	3
7		Workstations	109
8		Servers	9
9		Other	88
10	Vendors	Nortel Networks	3
11		Cisco Systems	1
12	Physical Links	Total	118
13		Serial	75
14		Ethernet	43
15	Other	Network Clouds	3
16		Configuration Utilities	8

(a)

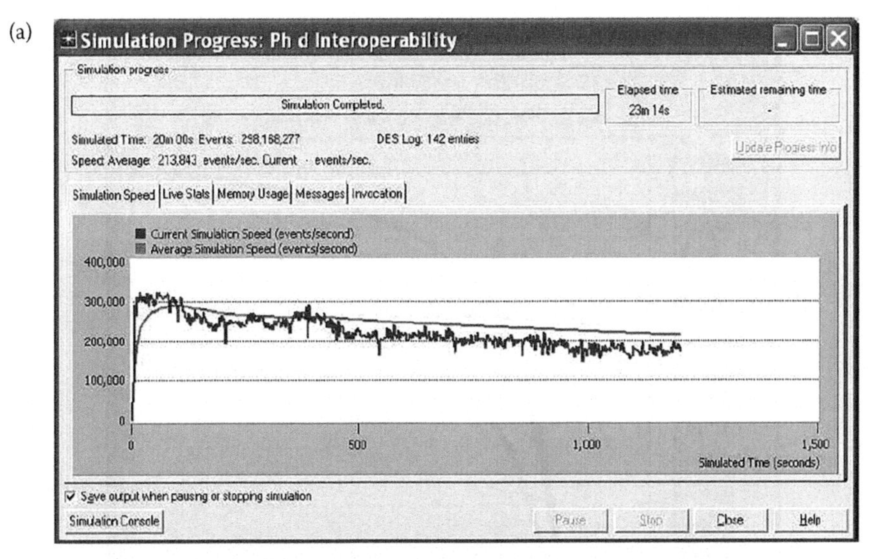

(b)

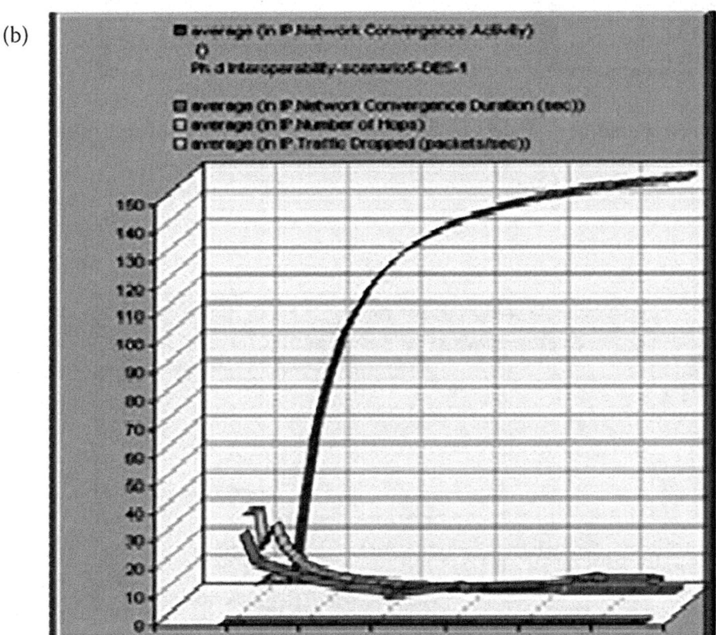

FIGURE 7.33 (a) 1 Simulation progress of an interoperability scenario. (b) IP network convergence—number of hops versus duration versus traffic dropped.

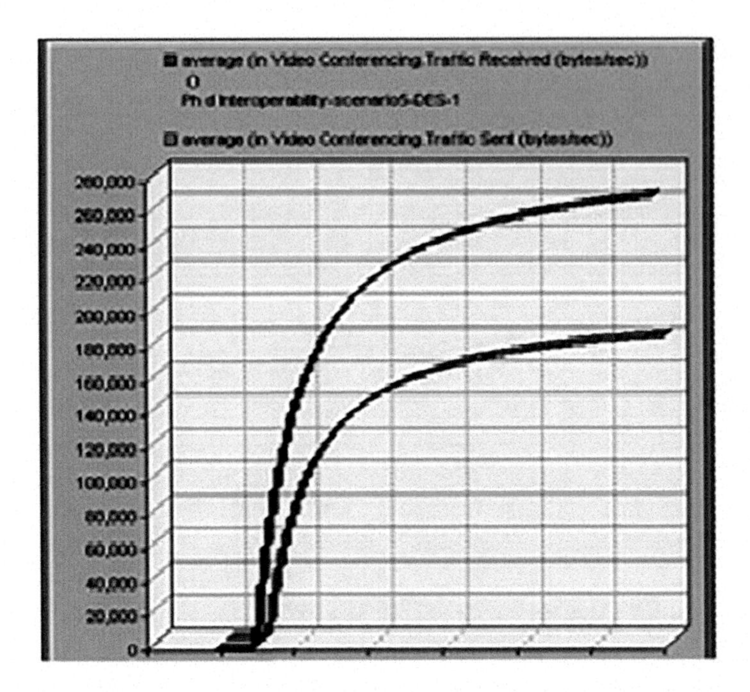

FIGURE 7.34 Interoperability—videoconferencing: traffic received versus traffic sent (bytes/sec).

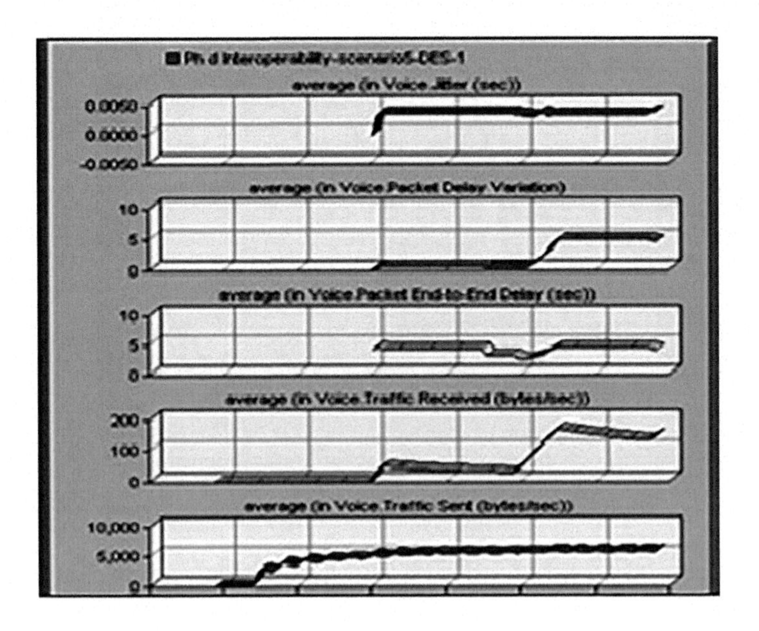

FIGURE 7.35 Interoperability—voice traffic.

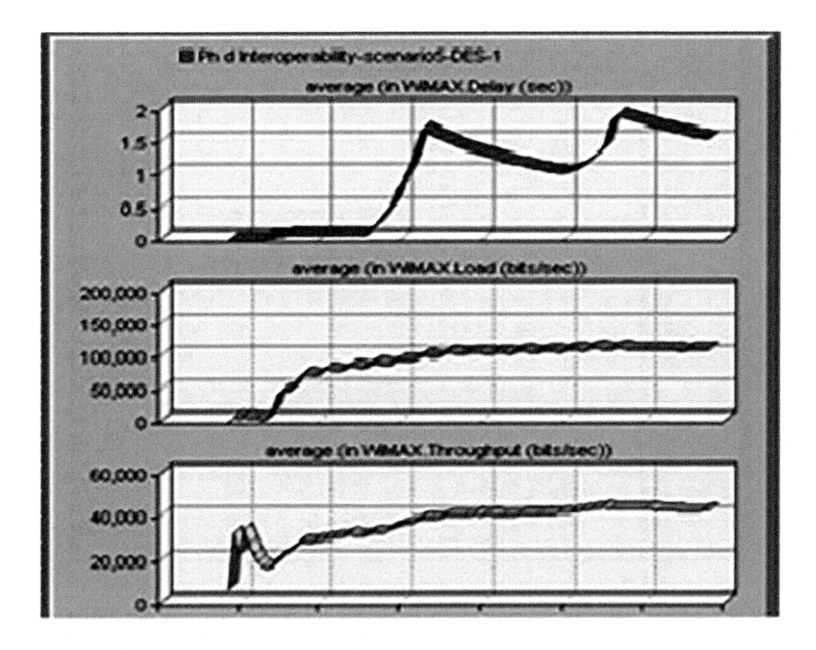

FIGURE 7.36 Interoperability—WiMAX delay versus load versus throughput.

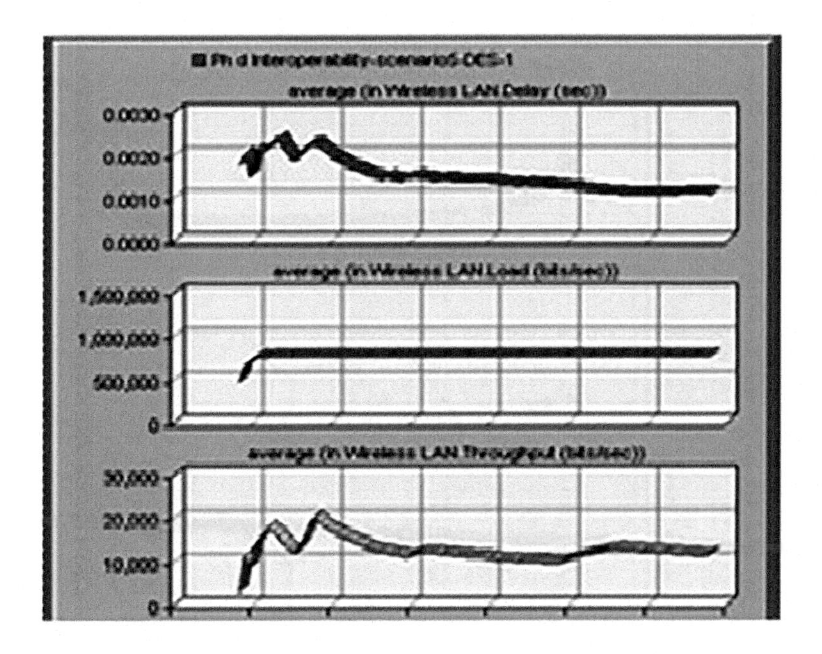

FIGURE 7.37 Interoperability—wireless LAN delay versus load versus throughput.

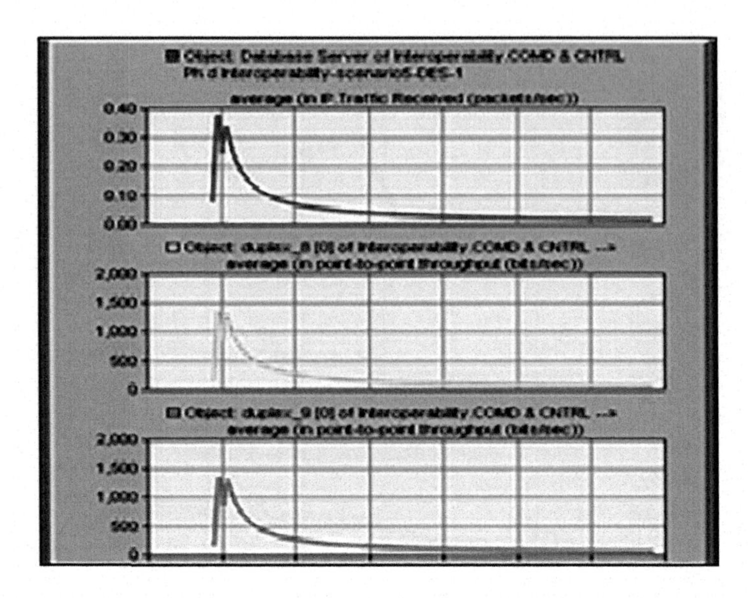

FIGURE 7.38 Interoperability—IP traffic received versus throughput for subnet COMD and CNTRL.

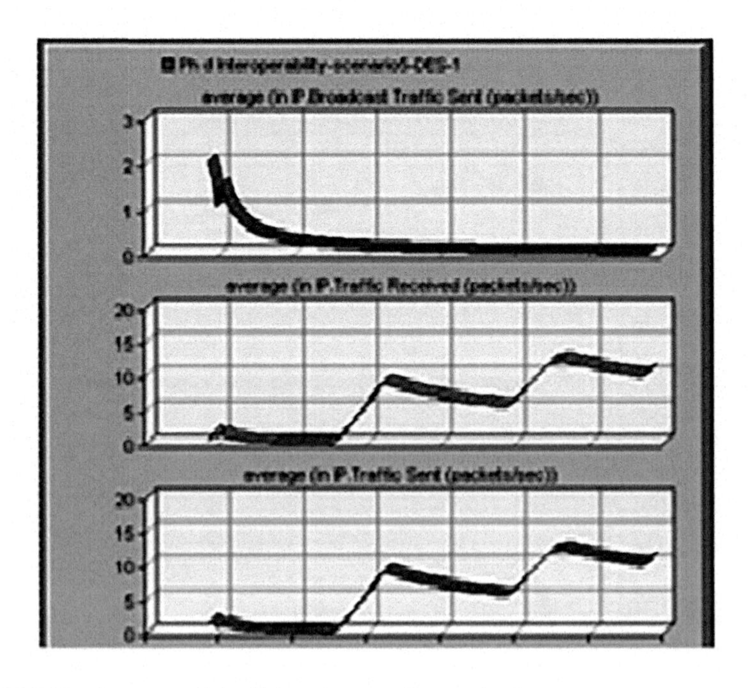

FIGURE 7.39 Interoperability IP broadcast traffic, received versus sent.

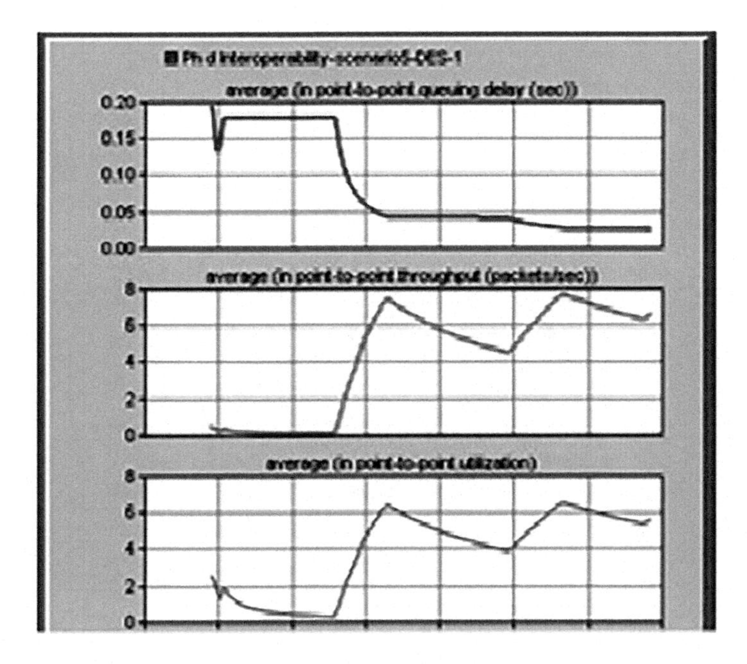

FIGURE 7.40 Interoperability—queuing delay, point-to-point throughput versus utilization.

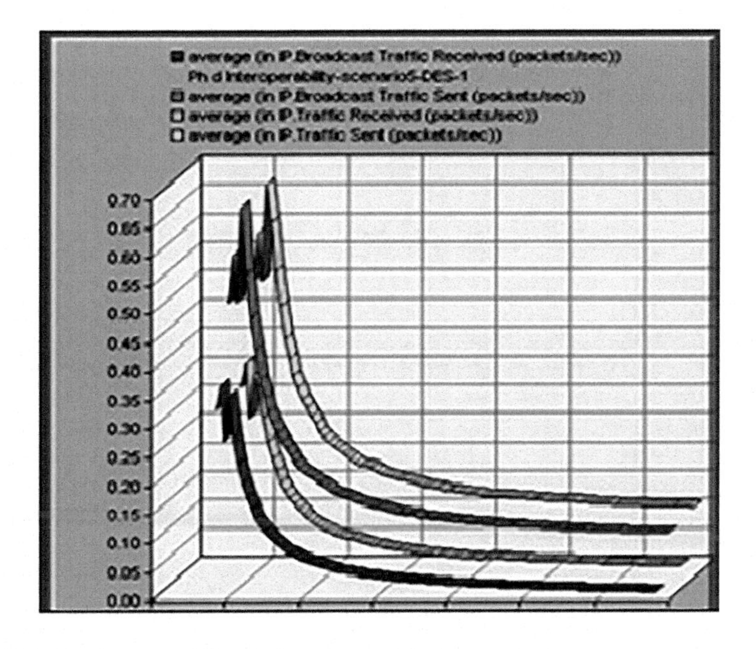

FIGURE 7.41 Interoperability IP broadcast traffic (packets/sec).

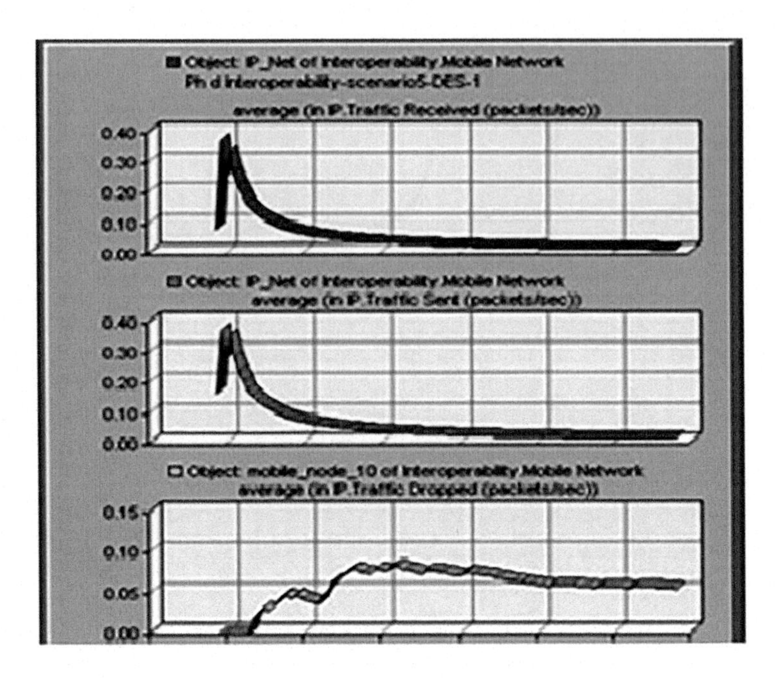

FIGURE 7.42 Mobile network IP traffic sent versus received versus dropped.

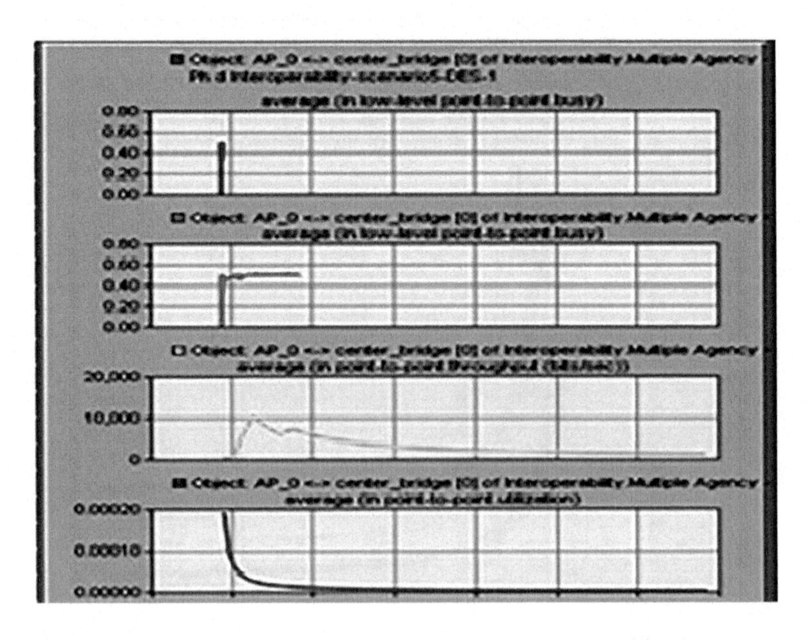

FIGURE 7.43 Multiple agency network—AP-0-throughput versus utilization.

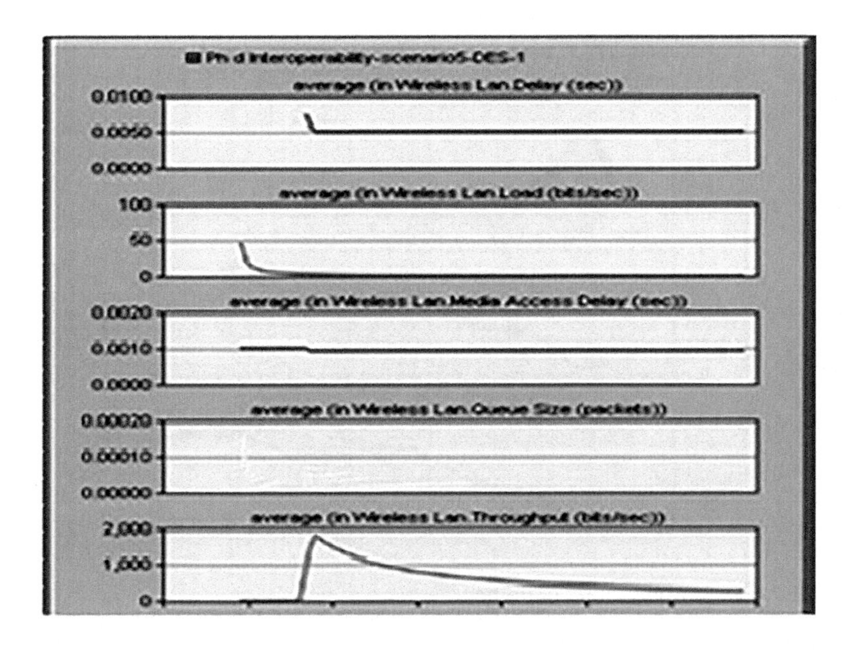

FIGURE 7.44 Interoperability wireless LAN results.

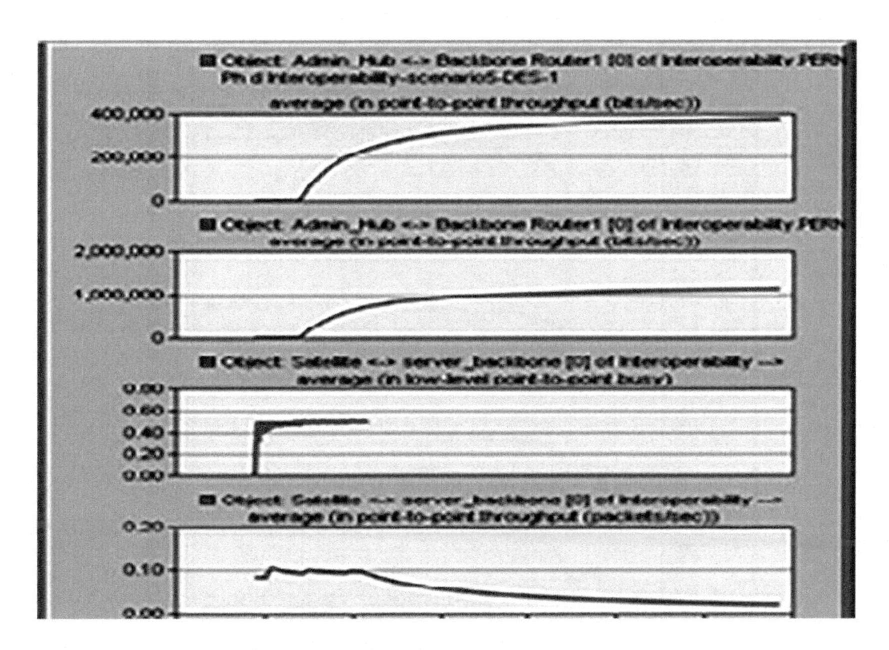

FIGURE 7.45 PERN/BBN router traffic and server backbone traffic.

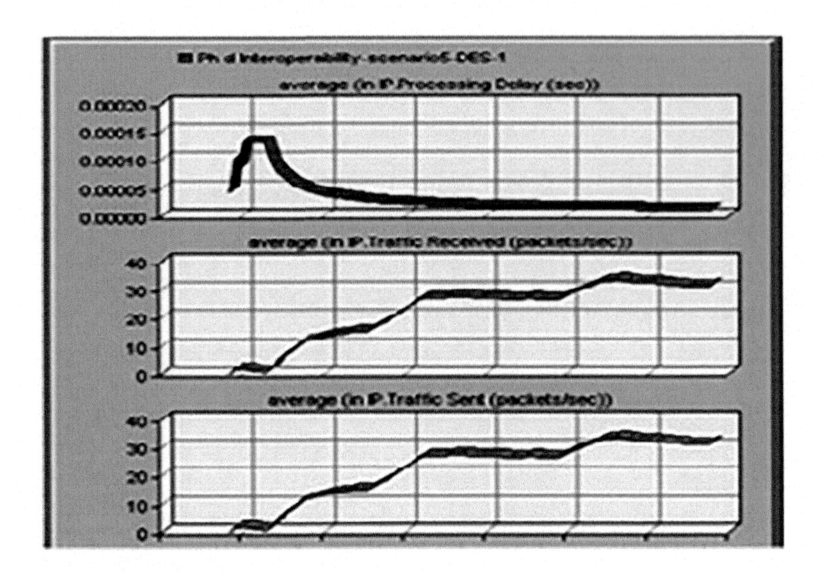

FIGURE 7.46 Interoperability IP traffic.

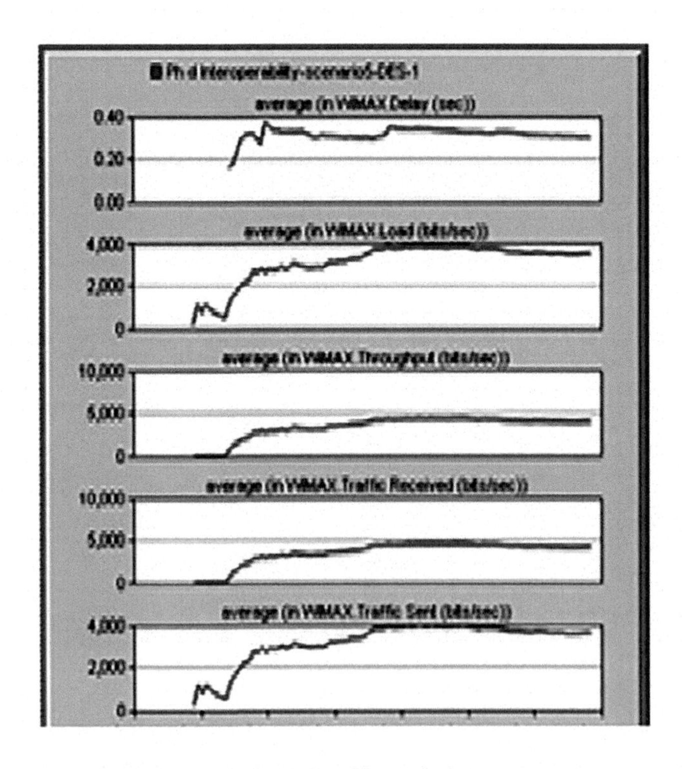

FIGURE 7.47 Interoperability—WiMAX traffic analysis.

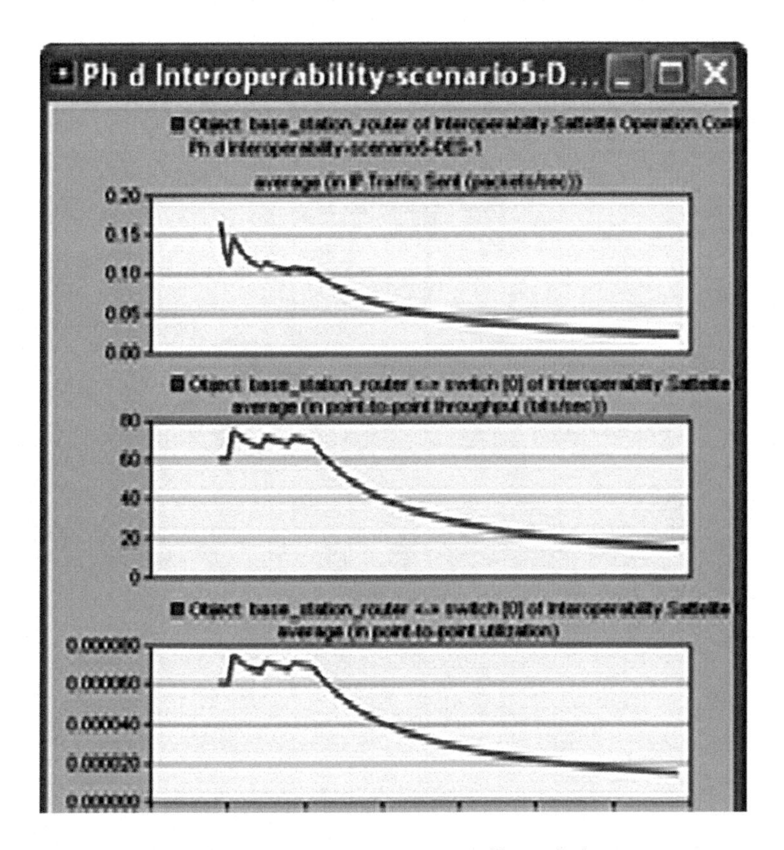

FIGURE 7.48 Interoperability—base station router traffic analysis.

7.7.4 Scenario III: Interoperability Scenario Linked with an Early Warning System

Two more scenarios are discussed in this section:

1. Interoperability scenario linked with the Early Warning Network
2. Interoperability scenario linked with the Early Warning Network—Zigbee Model

7.7.4.1 Interoperability Scenario Linked with the Early Warning Network: The Combined1 Model

This scenario assumes the role of the first responders' need of informing public, government, media and foreign country agencies of probable future threats of similar attacks of disasters during the response phase through early warning systems. The connectivity scenario is shown in Figure 7.49a,b. This scenario is termed as the Combined1 Model. The simulation progress of the Combined1 Model and the simulation results are presented in Figures 7.50 through 7.57 and Table 7.3.

(a)

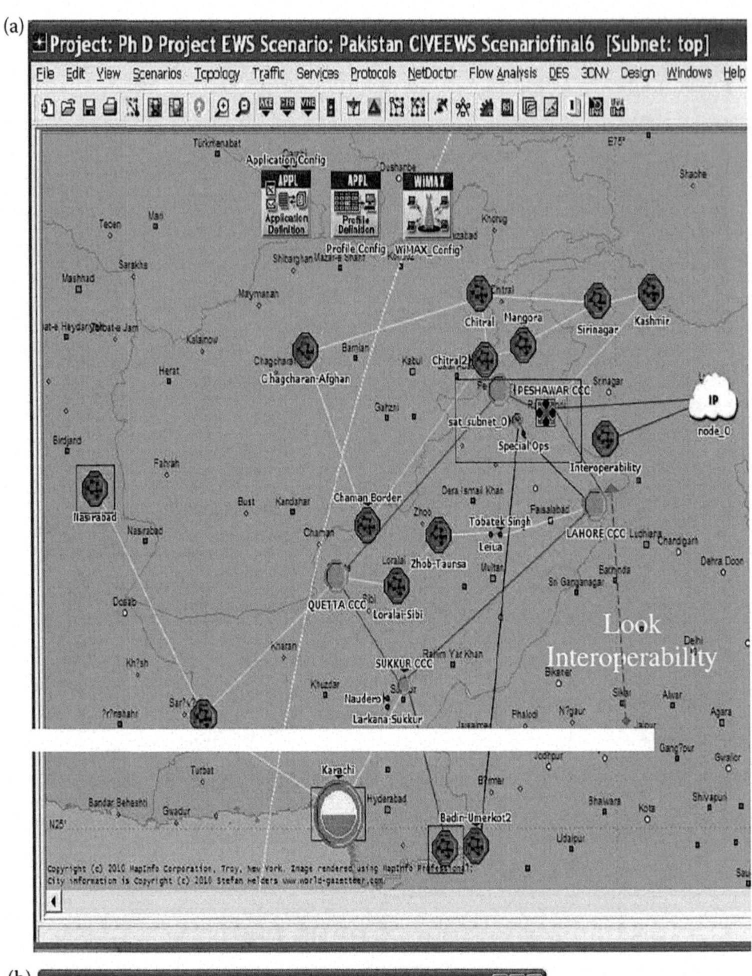

(b)

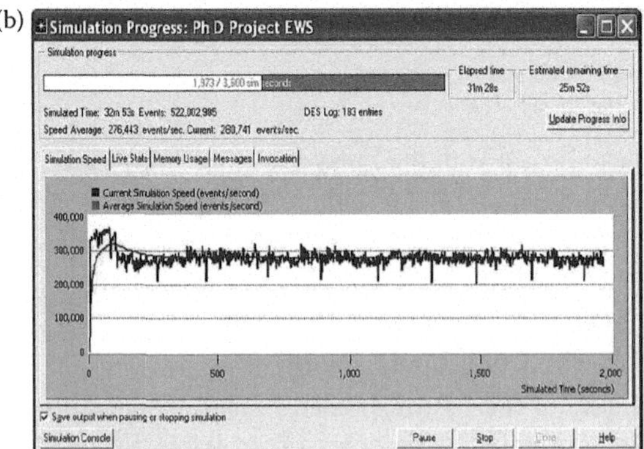

FIGURE 7.49 (a) Combined1 Model. (b) Simulation progress of the Combined1 Model shown in Figure 7.50.

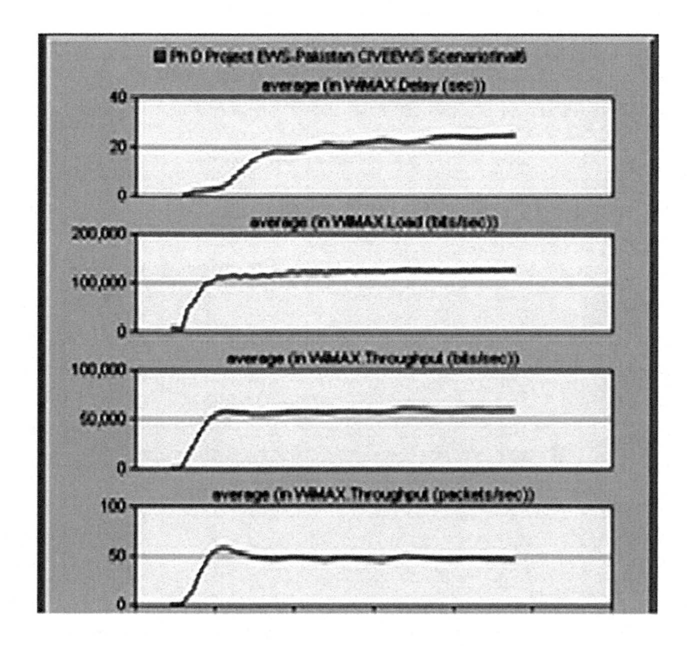

FIGURE 7.50 WiMAX results of the Combined1 Model.

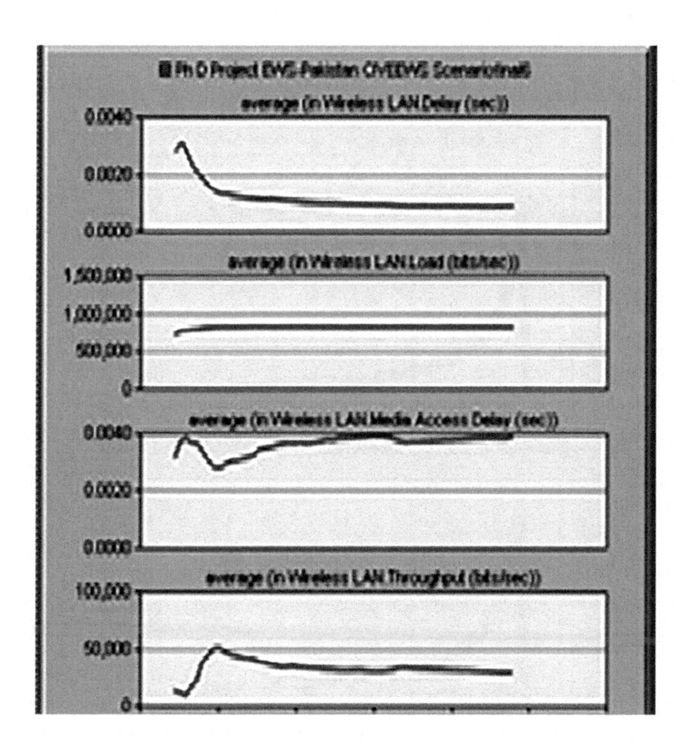

FIGURE 7.51 Wireless LAN results of the Combined1 Model.

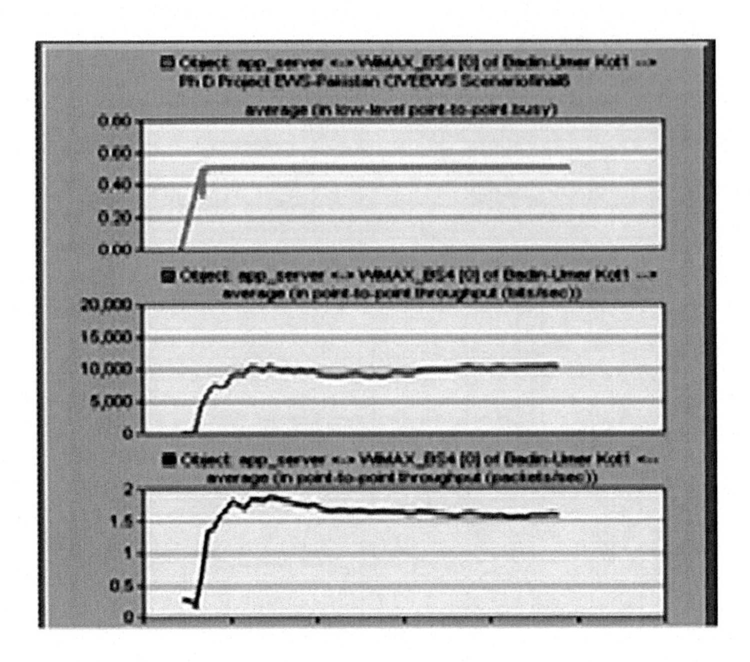

FIGURE 7.52 WiMAX base station 4 traffic results of the Combined1 Model.

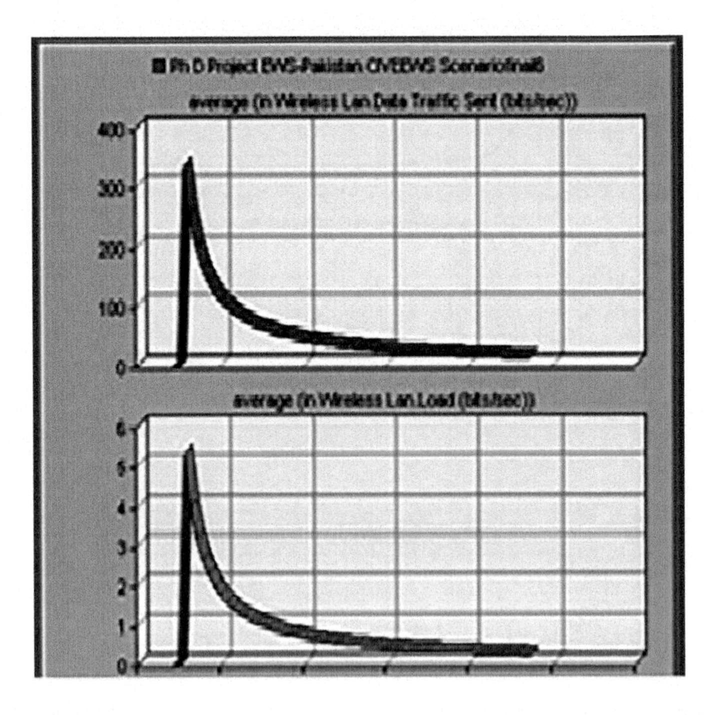

FIGURE 7.53 Wireless LAN load versus traffic sent results of Combined1 Model.

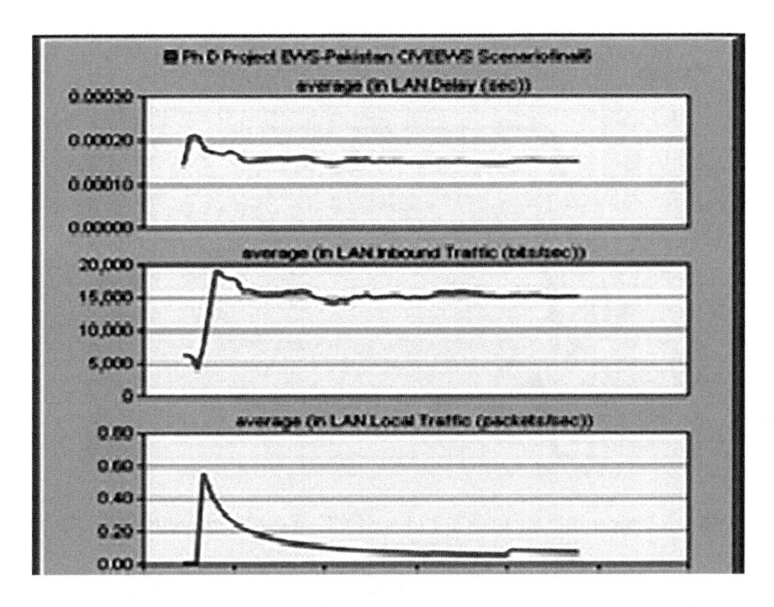

FIGURE 7.54 Wireless LAN delay versus inbound traffic versus local traffic in Combined1 Model.

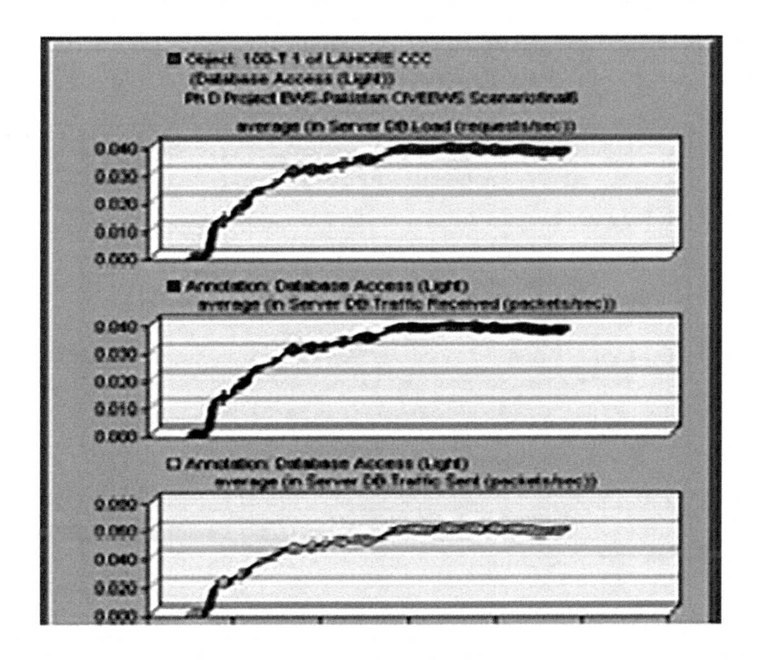

FIGURE 7.55 Annotation at Lahore CCC in the Combined1 Model.

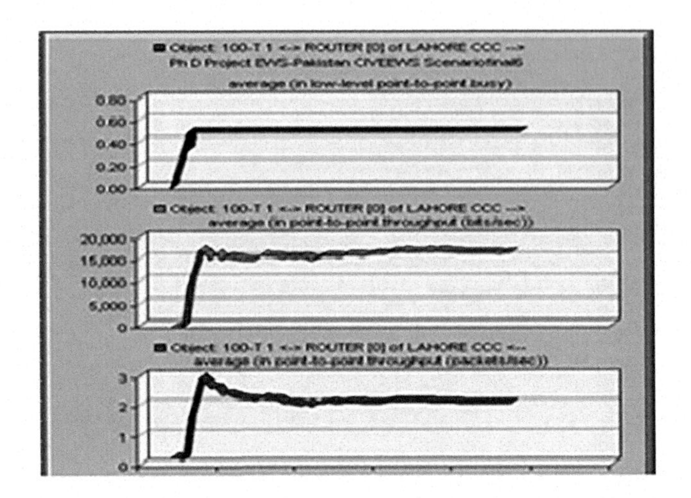

FIGURE 7.56 Router traffic analysis at Lahore CCC of the Combined1 Model.

7.7.4.2 Interoperability Scenario Linked with the Early Warning Network—Zigbee Model: The Combined2 Model

Figure 7.58 shows the Interoperability Model linked with the Early Warning Network—Zigbee Model. The scenario is termed the Combined2 Model. The role of this model is similar to the Combined1 Model but in a Zigbee-enabled mote environment.

7.7.4.2.1 Simulation Results of the Combined2 Model

The graphs in Figures 7.59 and 7.60 show a few simulation results from the Combined2 Model.

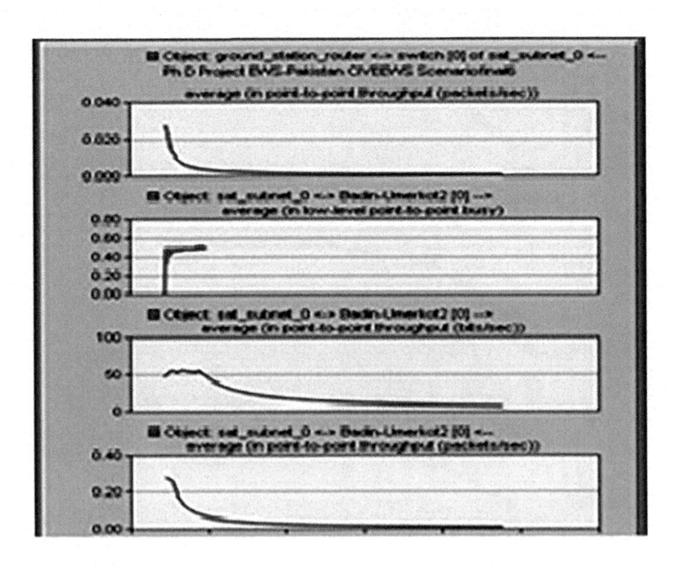

FIGURE 7.57 Traffic analysis of satellite subnets versus Badin subnet.

TABLE 7.3
Inventory Summary of the Combined1 Model

SNO	Element	Type	Count
1	Devices	Total	498
2		Routers	129
3		Switches	7
4		Layer-3 Switches	1
5		Hubs	5
6		LAN Nodes	13
7		Workstations	225
8		Servers	10
9		Other	88
10	Vendors	Alcatel-Lucent	5
11		Nortel Networks	3
12		Cisco Systems	1
13	Physical Links	Total	217
14		ATM	27
15		Serial	113
16		Ethernet	57
17	Other	Network Clouds	4
18		Configuration Utilities	71

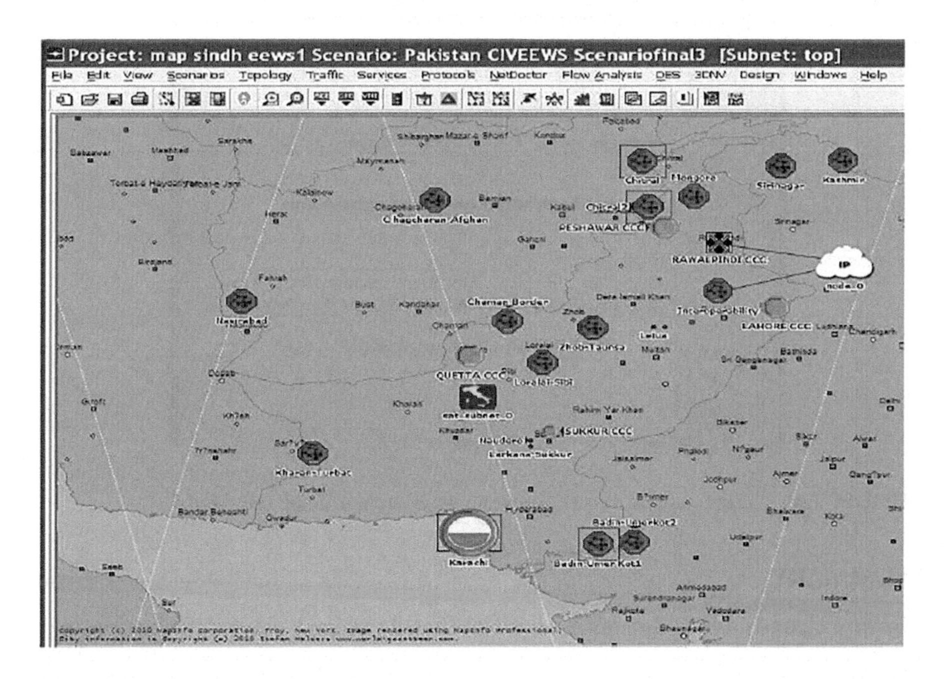

FIGURE 7.58 Interoperability model connected with early warning system in the Zigbee mode Combined2 Model.

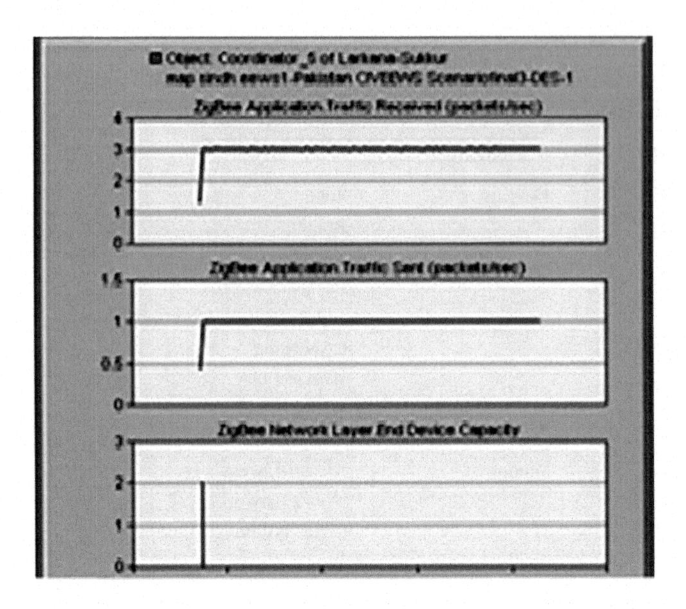

FIGURE 7.59 Zigbee MAC traffic analysis of the Combined2 Model.

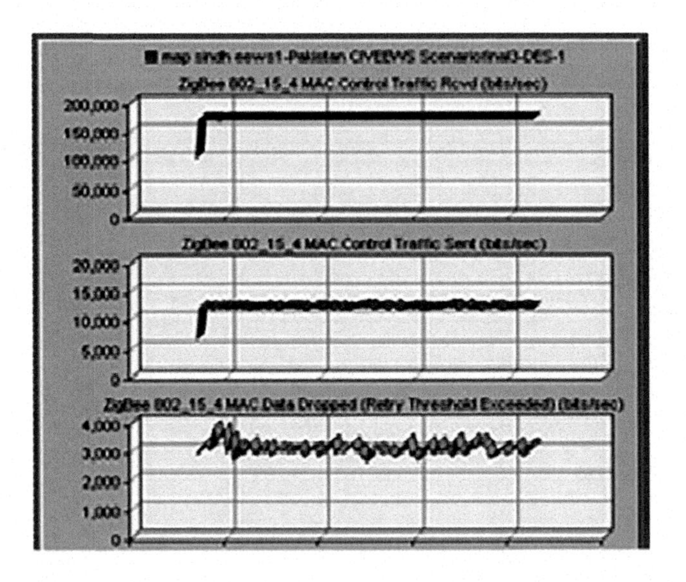

FIGURE 7.60 Zigbee application of traffic analysis of the Combined2 Model.

SUMMARY

In this chapter, communication networks are developed using information and communication technology for the multihazard disaster response phase. Various ICT models are developed, designed and implemented using the OPNET communication tool to cater to a real-time simulated environment. Results are produced and discussed for various developed models to meet the communication requirements of first responders; and timely warning messages and alerts for the community are discussed using various wireless technologies.

REFERENCES

1. Chris Oberg, J., Whitt, A.G., Mills, R.M., "Disasters Will Happen—Are You Ready?" *IEEE Communications Magazine*, January 2011: pp. 36–42.
2. Krock, R.E., "Lack of Emergency Recovery Planning—Is a Disaster Waiting to Happen?" *IEEE Communications Magazine*, January 2011: pp. 48–51.
3. Alzaghal, M.H., "Analysis of Jordan's Proposed Emergency Communication Interoperability Plan (JJECIP) For Disaster Response," Master's Thesis, Naval Postgraduate School, 2008.
4. M/S OPNET Inclusion USA. Available at www.opnet.com [Last accessed on January 19, 2010].
5. M/S Analytical Graphics. Available at www.agi.com [Last accessed on January 19, 2010].
6. S. ul Arfeen, S., Musavi, S.H.A., Shah, M.A., Kanwal, N., "Cross Domain Contour of AODV over OSPFv3 in Heterogeneous Ubiquitous Networks Using Internet Gateway," *Australian Journal of Basic and Applied Sciences*, Vol. 4, No. 9, 2001: pp. 4509–4521, ISSN 1991-8178.

8 Conclusion and Future Directions

INTRODUCTION

Managing vulnerability to natural and man-made disasters through modern information and communication technology is one of the scopes of this research work to achieve the desired objectives. The natural hazards, their vulnerability or both can cause failure of the technology, so engineers must design systems that will sustain the devastating impacts of the disasters. Since current design criteria are usually based on past experiences, we have proposed the ICT solution for a developing country like ours based on the basis of the practices and experiences of the developed countries. We have studied the programs and plans in place for disaster management in the United States, Japan, Italy and Turkey and then developed the ideas as invoked, in addition to a few novel ideas and innovative models. These pertain to designing a viable multihazard early warning system and designing an interoperable communications network for emergency response personnel, government agencies, NGOs and the public.

Disaster management can only be effective if its suggested infrastructure is stretched down to the Tehsil Level in Pakistan. Furthermore, the important political players and administrative heads of the four provinces and services chiefs along with the heads responsible for budgeting and financing, the NGOs and the volunteers should be a part of the disaster engineering and response community. Effective early warning systems for all hazards must be designed, developed and put into action throughout the geographical and territorial jurisdiction of Pakistan. These alerting systems should be linked through redundant communication links so as to provide backup facility in case of failure of one of them. Use of wired and wireless technologies with sufficient backup of battery power must be considered. The country should switch on to modern environmental sensor systems interlinked with early warning systems through wired, WiMAX, Wi-Fi, Zigbee and satellite-based networks. These networks must possess the capability to electronically connect first responders during the response phase with the Internet and technologists. The intra- and intercommunication of voice, text, geospatial and multimedia data transfer capabilities must be transformed by these networks. The roles of satellite communication, GIS, GPS, remote sensing, Zigbee and satellite phones must be embarked upon in communication networks owned by the disaster management agencies in Pakistan. The use of redundant emergency communication systems like terrestrial fixed services, satellite mobile phones, mobile and wireless services, ham radio (amateur services), television, radio, SMS, public warnings and notification systems, GIS, earth observatory satellite services, GSM, VoIP, software like the Sahana Disaster Management System, Internet emergency communication vehicles connected with satellites, remote sensing, videoconferencing and telemedicine devices, equipment, software and hardware has been emphasized in this book that is derived from my PhD thesis. The system of E-911 [1,2] calls, as is the practice in the United States, which is based on VoIP packet-switching technology, is recommended for Pakistan's 115 emergency services. This system has the capability to indicate geographic location on a map of the caller with the help of GIS- and GPS-enabled devices. We recommend a common platform for all agencies working for the cause of disaster resilience in Pakistan (Section 2.6) with interoperable communication networks and devices.

Due to the frequent occurrence of earthquakes, storms and floods, there is an urgent need to design and implement building codes for earthquakes and windstorms. Land-use planning and zoning, as well as its legislation to prevent people from seeking settlement alongside flood-prone areas, must be made.

Managing disasters should not be confused with averting them; rather, their impacts can be mitigated. Technology plays an important role in reducing the pre- and post-disaster havocs. The global population growth, climate changes, trend of poverty increase and hunger in developing countries like ours, changing political environments and rivalries and lack of administration, planning, good governance, misuse of administrative powers, malpractices, corruption and injustice are the core contributors in making the nations vulnerable to both man-made and natural disasters. Industrialization is continuously increasing with a greater percentage of both population and property that are monotonously under threats of disasters. It is thus concluded that technology alone cannot be the sole responsible factor for making the nations resilient to disasters. Instead, technology in concurrence with these contributing factors mentioned can only alleviate the vulnerability to the disasters. The potential of disaster risks can also be reduced through ideas and techniques invented and submitted by technologists, engineers and researchers. Hence, the technology introduced by these professionals must be paid due heed and consideration to mitigate impacts of disasters.

Furthermore, misapplication of technology and misguided actions on the part of first responders as well as by the population under threat must also be taken into account and thus avoided. So, there arises a forceful need to train for all disaster impacts, their potential and what actions should be taken for planning, preparedness and during the response and recovery phases. There appears to be no single organization, agency or institute providing disaster engineering training in Pakistan. So, there is a dire need of university, college or degree awarding institutes in Pakistan to offer courses at the undergraduate, graduate or higher levels in disaster engineering or disaster management in Pakistan.

8.1 RECOMMENDATIONS FOR THE FUTURE

1. A disaster engineering department and disaster engineering training institutes must be established and assigned the responsibility of the disaster management issue.
2. The disaster engineering infrastructure must be stretched down to Pakistan's Tehsil level.

3. Pakistan's satellite PAKSAT I or II must be dedicated to disaster management purposes.
4. Smart sensors like Zigbee-enabled motes are necessary to be deployed to record fire events and dispatch of incident information to communication headquarters so as to avoid manual call information regarding the occurrence of fire events.
5. High-resolution cameras with recording capability need to be installed in buildings with potential fire risk.
6. Computer-assisted systems are to be designed to manage, control and inform fire scenarios.
7. The CAD system must be electronically linked to the communication devices and networks of law enforcement agencies, firefighting personnel, hospitals, mobile emergency vehicles, traffic departments and water, fire and power departments. All these must possess communication devices that are interoperable and have the capability to broadcast and receive all types of data such as voice, high-quality images, SMS, Internet and videos.
8. The system of E-911 calls, as is the practice in the United States, which is based on VoIP packet-switching technology, is recommended for Pakistan's 115 emergency services. This system has the capability to indicate geographic location on a map of the caller with the help of GIS- and GPS-enabled devices.
9. Portable video cameras need to be mounted on the rooftops of vehicles of the police and fire brigades which transmit live movies of the disaster data to the receiving heads.
10. Bomb disposal squads must carry chemical, biological, radiological and secondary explosive search through respective sensor technology before clearing the scene.
11. Public safety wireless devices enabled with GIS, GPS facility and having the capability to connect to satellite, wired, wireless medium must be assigned to all first responders and operational heads for intra- and intercommunications among the networks.
12. Firefighter helmets must be mounted with cameras capable of transmitting live videos of the scene to their commanders.
13. Surveillance cameras with 24/7 storage ability must be installed in public places of strategic importance.
14. A National Criminal Database must be prepared, made publicly accessible, updated on a regular basis and linked to the National Database Registration Authority (NADRA). The vehicles registration database and stolen vehicles database must also be computerized and linked to investigators' offices.
15. The computerized database of weapons and licences of all types assigned to governments and civilians must be prepared and declared.
16. A Multihazard Early Warning System must be designed, developed and linked to all landline, wireless systems and with subscribers' mobile phones.
17. A disaster volunteer computerized database must be prepared. Proper training for these must be arranged. Volunteers from the government side

must be on a combined platform of agencies, such as the police, rangers, armed forces, Navy, Marine, CIA, NAB, courts, education, PIA, CAA, health, NADRA, SUPARCO, Civil Defence, NDMA, PNDMA, FFC, etc.

18. PDAs enabled with GPS-GIS technology and linked with high bandwidth radio and satellite channels must be given to first responders.
19. Satellite connectivity, VSAT terminals, satellite imagery and GIS must be given top priority among law enforcement agencies.
20. All government office records, records of the judiciary and police stations and records of civil infrastructure and hospitals must be computerized and linked to a one window operation.
21. Interoperable wireless networks for disaster management having the capability to connect local, national and international persons and systems must be designed and developed.
22. Systems should be made uncongested and accessible without failure from all communication devices and channels.
23. Building codes must be redesigned, publicly published and implemented with clear legislation measures. Structural specialists and civil engineers must be equipped with theodolites.
24. The number of government-owned ambulances must be increased; similarly, the number and quality of fire brigades must also be improved.
25. A computerized loss estimation tool must be designed and developed. The Sahana Disaster Management Tool should be implemented at all hierarchies of disaster engineering.
26. There should be a place called "Disaster City for Internally Displaced (DiCFID)" for large disaster victims. We recommend a province-wise allotment of lands for these cities. In Sindh, Nooriabad, near Karachi, is the best place. These places must be equipped with facilities of shelters, electricity, drinking water, communication and food. All relief operations should be focused at this place.
27. Disaster engineering institutes should launch undergraduate, graduate and masters courses in disaster technology. The Higher Education Commission and Pakistan Engineering Council should grant recognition and certification to these courses.

REFERENCES

1. Kahan, J.H., Allen, A. C., George, J. K., "An Operational Framework for Resilience," *Journal of Homeland Security and Emergency Management*, Vol. 6, No. 1, 2009.
2. Gupta, N.K., Dantu, R., Schulzrinne H., "Next Generation 9-1-1: Architecture and Challenges in Realizing an IP-Multimedia-Based Emergency Service," *Journal of Homeland Security and Emergency Management*, Vol. 7, No. 1, 2010.

Index